Engineering Specifications Writing Guide

Martin A. Fischer has practiced engineering for more than thirty years, including several years' experience in the design, analysis, and testing of a variety of systems and components for the Department of Defense, the Department of Energy, the commercial power industry, and other manufacturing concerns. He is currently a contract engineer whose responsibilities include writing specifications for replacement power plant equipment and for new Air Force equipment.

Martin A. Fischer

ENGINEERING SPECIFICATIONS WRITING GUIDE

An Authoritative Reference for Planning, Writing, and Administrating

A SPECTRUM BOOK

Prentice-Hall, Inc., Englewood Cliffs, New Jersey 07632

Library of Congress Cataloging in Publication Data

Fischer, Martin A.
  Engineering specifications writing guide.

  "A Spectrum Book."
  Includes index.
  1. Engineering—Contracts and specifications.
2. Specification writing.   I. Title.

TA180.F57   1983   808'.06662   83-4610
ISBN 0-13-279208-7
ISBN 0-13-279190-0 (pbk.)

10   9   8   7   6   5   4   3   2   1

ISBN 0-13-279208-7

ISBN 0-13-279190-0 (PBK.)

Editorial/production supervision by Peter Jordan
Manufacturing buyer: Cathie Lenard
Cover design © 1983 by Jeannette Jacobs

This book is available at a special discount when ordered in
bulk quantities. Contact Prentice-Hall, Inc., General
Publishing Division, Special Sales, Englewood Cliffs, N.J. 07632.

Prentice-Hall International Inc., *London*
Prentice-Hall of Australia Pty. Limited, *Sydney*
Prentice-Hall Canada Inc., *Toronto*
Prentice-Hall of India Private Limited, *New Delhi*
Prentice-Hall of Japan, Inc., *Tokyo*
Prenctice-Hall of Southeast Asia Pte. Ltd., *Singapore*
Whitehall Books Limited, *Wellington, New Zealand*
Editora Prentice-Hall do Brasil Ltda., *Rio de Janeiro*

To my beloved parents

Contents

The Flow of Requirements 13

SPECIFICATION MACROSTRUCTURE

Types of Engineering Specifications 21

Specifications Precedence and Referencing 26

Component Specification Structure 30

System Specification Structure 59

SPECIFICATION MICROSTRUCTURE

PREPARING SPECIFICATIONS

Preface

This book presents a comprehensive set of guidelines and instructions for preparing engineering specifications—particularly for those specifications sent out, with or without drawings, to obtain bids for and procure equipment. Since most such documents are part of a project specification hierarchy, the organization of such hierarchies is also given consideration. As a result, system design, system test, component design, component purchase, and component test specifications are explicitly covered.

The material is divided into four parts: Specifications in Perspective; Specification Macrostructure; Specification Microstructure; and Preparing Specifications. The first part is concerned with how specifications fit into procurement documentation and project technical documentation. The second part is concerned with overall specification structure at three levels. The third is concerned with details of material arrangement within these levels. Part IV is concerned with the work involved in preparing a specification for release to its users—that is, getting it "out the door." This part also includes a discussion of the initial procurement, using the specification, wherein those who may have to work to it have their say concerning its contents. Finally, there is a set of appendixes consisting of detailed topic outlines and checklists. Once specification writers, reviewers, and editors are familiar with the process, they may use these appendixes in their day-to-day work without continually refering to the first four parts.

This book occupies a place alongside the few current volumes on construction (building) specifications, Military Standard 490, entitled *Specification Practices*, and the Department of Defense manual 4120.3-M, entitled *Standardization Policies, Procedures and Instructions* (containing a section equivalent to MIL-STD-490 without its appendixes). These last two documents contain the only published treatments of engineering specifications, as distinct from construction specifications, in wide use. Unfortunately, they have a very strong bias toward military equipment and its requirements. As a matter of interest, these classes of specifications are discussed briefly in Part I.

I wish to express my appreciation to Russ Stultz for his most thorough review of my initial manuscript and for his many useful comments. He found the weak spots. I also am grateful to my manuscript typist, Helen Mosbrook, for her effort in preparing that first revision.

Engineering Specifications Writing Guide

Part I

Specifications in Perspective

Chapter 1

Procurements and Specifications

1.1 PROCUREMENT DOCUMENTS

In industry and government today, no procurement of goods or services is accomplished without some form of documentation. The procurement of goods takes more extensive documentation than that of services, and when these goods are what is known variously as equipment items, engineered items, or, more commonly, hardware, the documentation involved may become very complex. However, in spite of the seeming complexity, each document is one of three major types: contracts, administrative specifications, and engineering specifications. These are shown in Figure 1.1, which also indicates how the types of documents relate to one another.

1.2 CONTRACTS

Contracts are familiar legal documents, with one contract per procurement. Large long-term projects may be an exception. Often, these projects are divided into phases, with a separate contract for each phase.

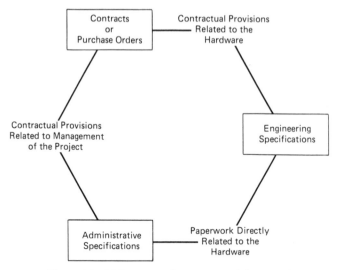

Figure 1.1 Major types of procurement documents

For example, procurement of a new engine control may be divided into three phases: (1) a study phase, culminating in the delivery of a design report, (2) a prototype development phase, culminating in the delivery of a set of drawings and a test report, and (3) a production phase for a given quantity of controls.

Contracts contain all of the pertinent legal provisions or requirements, enforceable under the law. Contracts are the controling documents basically concerned with cost and delivery. If necessary, contracts can modify the engineering and administrative specifications they impose on those contracted to do the work. A purchase order replaces the contract for a similar procurement and performs the same functions as a contract. People who write contracts cannot change the laws that apply; they can only follow them and interpret them as they affect different procurements. Basic contractual forms and requirements are the province of legal departments. Once these have been determined, individual contracts are prepared, frequently by purchasing departments.

1.3 ENGINEERING SPECIFICATIONS

Engineering specifications contain all of the requirements directly applicable to the hardware being procured, and also the requirements affecting the hardware during delivery and use. Included in the latter are all supporting written materials, often called *related software*.

4

Engineering specifications define the desired end result of a procurement; therefore, they come second in the contractual hierarchy. People who write engineering specifications work primarily in accordance with natural laws and with considerations derived from natural laws. They cannot change these laws or appeal them. The consequences of violating these laws are irrevocable. Engineering specifications are, of course, the province of engineering departments.

1.4 ADMINISTRATIVE SPECIFICATIONS

Administrative specifications contain all requirements that are not included in contracts or engineering specifications. That is, they contain requirements not directly applicable to the hardware but to the associated administrative process of procuring it, and general requirements for the hardware-related software. These include requirements for planning, scheduling, cost accounting, reporting, document submittals, procedures to be followed, and forms to be used. Because administrative specifications are meant to facilitate production and delivery of the desired products in accordance with the contract, they take third place in the contractual hierarchy. People who write administrative specifications work in accordance with logical procedures that are under their own control. They are free to modify these procedures as they see fit to achieve a more effective approach. Administrative specifications are the province of purchasing departments and project or contract administrators.

1.5 INTERCONNECTIONS

The connections between a contract and its engineering specifications are in the areas of meeting requirements for durability (warranties), safety (liabilities), necessary information (disclosures), use of codes and standards (regulation and inspection), and expert assistance (field service). Engineering specifications define what is expected of and for the equipment, and contracts define related conditions and limitations. Engineering requirements and legal requirements must be arranged in a form that is mutually compatible, even though the contract is the controlling document. Finally, care must be taken to assure that engineering requirements go into the engineering specifications and that related legal requirements go into the contract. The latter should not appear in the engineering specification in a disguised or semidisguised form.

The connections between engineering and administrative specifications are in the areas of paperwork directly related to the hardware (that is, reports concerning design, manufacturing, testing, procedures, records, instructions, manuals, and the like). Generally speaking, engineering specifications define what is to be submitted, and administrative specifications define how and when it shall be submitted, including requirements for transmittal forms, certifications, approval schedules, and so forth. The form of administrative requirements must not be incompatible with that of engineering requirements, otherwise the goal of facilitating a procurement will not be achieved.

The connections between contracts and administrative specifications are in the areas of contractual provisions related to management of the project and the terms of payment. Contracts define what is to be delivered and when, and administrative specifications define how delivery will be made, how it will be verified as acceptable, and how payment will be authorized. The contract defines the penalties to be imposed if its terms are not met, and the administrative specification defines procedures for failure assessment. The form of contractual requirements ultimately determines the form of associated administrative requirements. Contracts and administrative specifications have more in common with each other than either does with engineering specifications. The first two are based on man-made laws and rules (external to the writing group in the first instance, and internal to it in the second), and the last are based on natural laws. Much interdepartmental misunderstanding and conflict arises from the failure to recognize this fact.

Chapter 2

Engineering Specifications

2.1 CLASSES OF SPECIFICATIONS

Three broad classes of specifications have evolved from the development of standardized specification-writing practices. In chronological order, they are construction specifications, military and federal specifications, and engineering or equipment specifications. All have their origins in military or government-directed work.

Construction specifications were standardized during World War II, when the Army Corps of Engineers needed to achieve more effective control of the vast number of complex construction projects it had under way. Now their form is determined by the Construction Specifications Institute. Military Standard 490 on Specification practices appeared after World War II, when the Cold War made it clear that military development projects were going to become a continuous, long-term undertaking. Much of this standard has now been incorporated into the Department of Defense Standardization Manual. The first attempts at establishing other standards for engineering specifications occurred within the Atomic Energy Commission when it became evident that writing practices that were suitable for field construction sites were *not* suitable for engineering office, shop, and testing laboratory work. Furthermore, military practices were too strongly oriented toward mobile weapon systems. Unfortunately, the initial work was not carried to completion and no document similar to books on construction specifications or Military Standard 490 was published.

The top level of Figure 2.1 shows where each class of specifications originates with the ultimate users, sellers, or distributors of the specified end product. At this point, reference to the glossary (page 121) may be helpful. The definitions contained therein are important for a clear understanding of much of the following material.

Military specifications (block 1 of Figure 2.1) originate within the services and usually start out with weapon system specifications. Most *federal specifications* originate in the General Services Administration and are concerned with materials requirements or the products that will stock federal warehouses all over the country—specifications for pencils, paper, soap, towels, garment materials, tires, motor oil, grease, paints, cleaning fluids, and much more.

Industry and nonmilitary government specifications (block 2) originate within private organizations and government organizations such as the Department of Energy. They usually start out with some overall design description, giving functions and limits for the envisioned project. An example would be a new chemical processing system for quantity production of a plastic recently developed in the laboratory.

Facilities are required by industry and government organizations, both military and nonmilitary, for housing systems and the people who operate and service them. This brings about definitions of facility requirements (block 3), which complement the system requirements. In many instances these facility requirements may describe the expansion or modification of an existing facility to accommodate the new system. Here, then, are the origins of construction specifications.

If taxpayers and consumers are not brought into the picture, these ultimate users at the top level are also the ultimate sources of funds. Their main functions are definition of the product and its acceptance criteria, project management, and disbursement of funds.

2.2 LEVELS OF SPECIFICATIONS

The second level down in Figure 2.1 is the top level of producers or suppliers of the final product. In the military area (block 4) there is a prime contractor or several associate contractors. These organizations

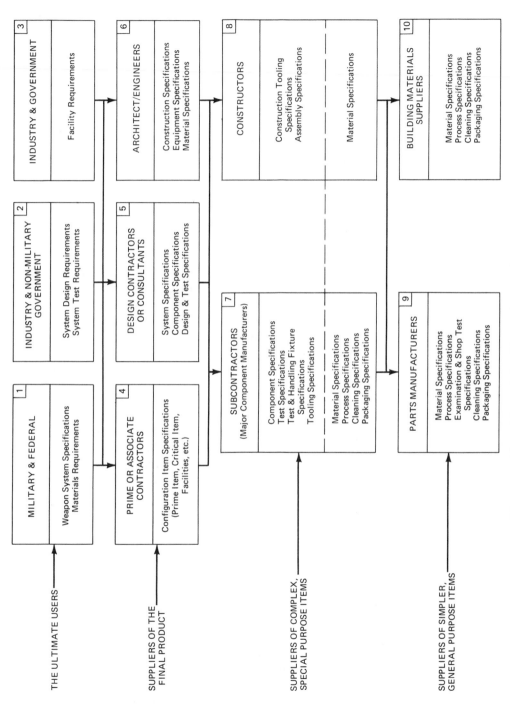

Figure 2.1 The origins, relationships, and types of specifications

write specifications for their own use and for use by their subcontractors. Specifications originating here are for prime items, critical items, facilities, and so on. The term *configuration item* usually refers to a subsystem of the weapon system, separately deliverable, and these items are further designated prime, critical, or noncritical. All of these specifications are written in accordance with Military Standard 490. Many items do not meet the criteria for the above classifications, and for such items the contractors are permitted to use something other than Military Standard 490.

Commercial organizations, including ones who also serve the military, function as design contractors or consultants to nonmilitary government agencies and other industrial firms (block 5). Depending upon the nature of the project, the terms of their contract, and their own intentions for subcontracting, they will write a variety of specifications. These may include system specifications, component specifications, design specifications, or test specifications, all of which are in the class of *engineering specifications*. Depending upon their origins and interests, these firms may also do manufacturing and testing. Hence, as on military projects, they will write specifications for internal use as well as for subcontracted work.

The detailed work of designing facilities, especially major ones, is done by those organizations known as architect/engineers. At these firms (block 6), most construction specifications are written. In order to do a complete job, they also write specifications for equipment and materials. These specifications are used for purchasing items shipped to the building site, while construction specifications govern work at the site. Architect/engineers may or may not handle their own construction work. If construction is subcontracted and the constructor is allowed to purchase materials, the construction specifications must contain many of the same type of requirements found in equipment and materials specifications.

At the third level down in Figure 2.1 the importance of manufacturing increases, and firms that only do engineering paperwork are no longer evident. This is the level of subcontractors (block 7), who supply major components to all those concerned with assembling the final product. Often, these components are of a complex or special-purpose nature. This is also the level of constructors (block 8), who do less engineering and more field work than most of the architect/engineers.

The fourth level down in Figure 2.1 is where manufacturing concerns predominate. Here are the suppliers of vast quantities of simpler general-purpose items. Parts manufacturers (block 9) supply such items as bolts, nuts, washers, wire, terminals, resistors, capacitors, and simple assembled items. Most of what they supply goes into numerous forms of equipment. Building materials suppliers (block 10) supply the construction materials for facilities housing the equipment and systems. Items such as lighting fixtures and washroom fixtures, which make the facility habitable by an operating crew, are usually included here.

The dashed line in Figure 2.1 divides the third level into two parts. Specifications below this line bear a strong resemblance to procedures. Their formats and standards are largely determined by various industrial and technical societies, associations, and institutes. Also found in this area is that vast body of federal specifications, mentioned earlier, used by the General Services Administration. A common practice is to have a specification written by the supplier of the item, subject to approval by the agency involved.

The vast majority of engineering and equipment specifications are written at the second level of this figure, or at the third level above the dashed line. This is also where most construction specifications and most military specifications following Military Standard 490 are found. This is the area covered by this book, particularly blocks 5 and 7.

The four levels of this figure have been selected arbitrarily for illustrative purposes. In reality, there may be as few as three or as many as five or six levels. The expansion would be at level three, where the distinction between complex, special-purpose items and simple, general-purpose items is relative. The contents of the midlevel blocks are also somewhat arbitrary, since the same large organization may function at more than one level. At any level below the first it is customary for an organization to refer to the next lower level as subcontractors, and any level further down as subtier contractors.

2.3 THE DISTRIBUTION OF REQUIREMENTS

The grand total of all requirements determining what the final product will do and under what conditions are not found in any one document. These requirements are distributed through all levels in all of the specifications for each definable item used to make the product. Requirements for the final product, determined by the ultimate user and specified at the top, are the starting point. Directly or indirectly, they determine the rest. Thus, there is a downward flow of requirements and what might be described as a settling-out at each level. There is also a continuing increase in requirements, with the total number of different requirements at any level exceeding the total for the previous higher level.

This ongoing distribution of requirements down the levels continually reaches points where words no longer suffice. At these points requirements (or the meeting of requirements) produce drawings that give pictorial form to previous literal forms. These drawings may represent a complete fulfillment of some requirements and only a partial fulfillment of others. An understanding of this process is necessary not only for good specification writing, but also for good project management.

Chapter 3

The Flow
of Requirements

3.1 THE NATURE OF REQUIREMENTS

The process of distributing requirements, the flowdown, and giving the hardware its form is best understood in the context of two questions: "What?" and "How?" In general, requirements answer the first question, and drawings answer the second. Requirements state what is wanted, what is to be accomplished. Drawings show how it will be done, the form the hardware will take. If a group of requirements forms the statement of a design problem, then the drawing or drawing package meeting those requirements is the design solution.

In a modified context, the *how* of one level becomes the *what* for the next lower level. The requirements of a specification sent to a design engineering group are the *what*, and the drawings they produce are the *how*. These drawings are sent to the shop, where they are now the *what* (what is wanted or what must be built), and the shop procedures are the *how* (how the item on the drawings will be built). This alternation between *what* and *how* is readily seen at the lower levels, where component specifications, drawings, and shop procedures are most in evidence. It is not as easy to see this process at the higher levels, where specifications generate more specifications. However, a closer examination of the process at these levels indicates that, in general the *how-what* description holds true. System requirements describe what the system must do. Subsystem requirements determine how each subsystem will contribute and what its components must do. Component requirements determine how each component must function to meet a requirement of its subsystem and what its subcomponents and parts must do. Thus, in some manner, the alternation between *what* and *how* occurs at all levels.

It follows that middle-level specifications imply a partial design solution, even though they define requirements only, for a particular item. The very breakdown into subsystems and components is an implied design solution for the total system. This is unavoidable. The overall requirements cannot continue down through the levels in their original form. They must be translated or recast into more meaningful or suitable forms, as determined by the breakdown chosen.

This discussion may prompt a question. The contract determines who will do the job, when it will be finished, and where the product will be delivered. The engineering specifications determine what will be done and how, as we go down through the levels. Shouldn't the specification also tell *why*? Shouldn't there be a triad of *what, how,* and *why*? The answer is *no*. The *why* may be discussed during reviews and contract negotiations, but once these are over explanations and justifications are no longer necessary. Specifications and drawings only define the *what* and the *how*; the *why* has previously been settled.

3.2 NONINTERACTIVE COMPONENTS

The flowdown of design requirements for noninteractive components is shown in Figure 3.1. When there are no interactions between components,

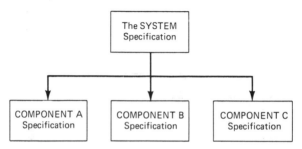

Figure 3.1 Flowdown of design requirements for noninteractive components

as discussed in the glossary, the flowdown is on a direct basis, often one-to-one. Every requirement in the upper-level specifications shows up at least once in the next level down, in its original form or a modified but recognizable one. Any additional requirements that are necessary at the lower level for completeness are simply added in. This makes tracing of requirements from top to bottom, and from bottom to top, a simple matter.

3.3 INTERACTIVE COMPONENTS

The flowdown of design requirements for interactive components is shown in Figure 3.2. Interactions between components, as discussed in the glossary, complicate the situation. Some requirements may flow

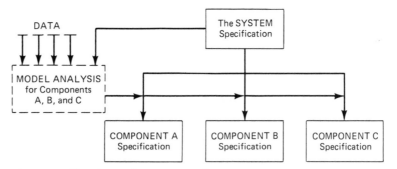

Figure 3.2   Flowdown of requirements for interactive components

down directly, but others cannot. A model must be developed using the interacting components and other data concerning the subsystem or higher-level system containing the components. A model analysis then supplies many necessary additional requirements for the components of the subsystem in the model. These requirements may bear little resemblance to the original system requirements. They are the inevitable result of mathematical methods, physical principles, and material properties as applied in the particular model of the system.

This situation creates a problem for tracing requirements to make sure that all are in the right places and are indeed necessary. If the additional data used are internal to a computer program, they may be very hard to identify. The solution for this problem is shown in Figure 3.3. Additional data needed for the analysis are recorded in a data book, where they can be kept current for a developing project and made available to other users. All results of the model analysis are recorded in a report. From this report necessary additional requirements are

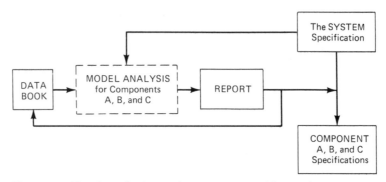

Figure 3.3 Flowdown for interactive components with records

distributed to the component specifications. The report can also be used as a source for additional information, which will be required by others, placed in the data book.

3.4 THE SPECIFICATION TREE

Figure 3.4 shows the specification structure and flowdown of requirements through the levels for a project. It also shows the appearance of drawings and their relationship to the specifications. At this point a reading of the first section of the appendix, Specification-Related Drawings, may prove helpful.

As Figure 3.4 shows, the main flow of requirements is from the subsystem design specification down through the major component design specifications and the component specifications to the subcomponent specifications. At the bottom level, just below the last level of specifications, are the final manufacturing drawings and data sheets for standard components and parts. At the higher levels, there are branch flows to the model analyses and back to the main line. There are also branches where the specialized design requirements have determined related test requirements.

The figure also presents an interchange of information between subsystem-level and component-level data books. In theory, information could go from the subsystem-level data book to the component-level model analyses, and from the component-level model study report back to the subsystem-level data book. New projects usually operate in a state of flux for some time. Changes occur, model analyses must be rerun, reports must be revised, and official approval for the use of new results must be obtained. If the flow of information is direct, confusion often arises over what information is being used at a particular moment

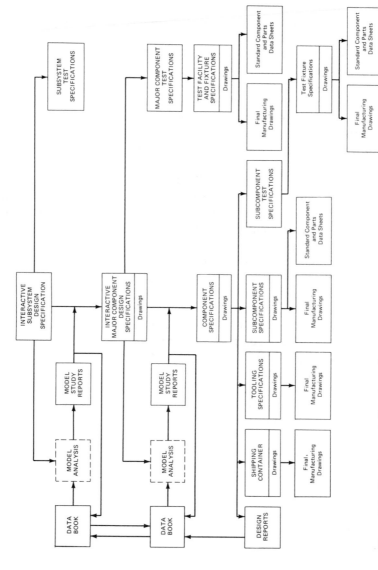

Figure 3.4 Requirements flowdown for a project

and why. Sometimes the component-level work gets ahead of the sub-system-level work, producing results that necessitate changes in the subsystem. By requiring the exchange to go through the component-level data book, a time lag is added, providing a stabilizing effect.

At the lowest level of specifications, requirements for specialized supporting specifications show up, such as for tooling, test fixture, and shipping container specifications. Detailed design reports appear here, which may provide still more information for the data books.

If you delete the data books, analyses, reports, drawings, and data sheets from Figure 3.4, what is left is a chart commonly called a *specification tree*. It shows the arrangement of and connections between all of the project specifications, and it looks like an expanded version of Figure 3.1. Specification trees are useful tools for determin-ing the specification-writing workload and for following up the writing process as a part of the total effort. Also, they are often considered a tool for verifying the traceability of requirements. This works reasona-bly well when everything in the system is noninteractive. However, when a number of interactive items enter the picture, complete tracea-bility of requirements presupposes a knowledge of where the data books, analyses, and reports enter the picture. A complete chart, such as Figure 3.4, is therefore necessary.

Although five levels of documentation are shown in Figure 3.4, only four or even three levels of organizations may be involved. The figure also shows interactions ending after the second level down. This, of course, is an arbitrary selection, just as the figure is an arbitrary example. For very complex subsystems several models may be neces-sary for groups of functional requirements, with a report issued for each group analysis. The functions and design of each item determine the occurrence and location of interactions. This figure also assumes that the subsystem can be tested, along with its interacting subsystems, without a special test facility.

Part II

Specification Macrostructure

Chapter 4

Types of Engineering Specifications

4.1 PRELIMINARY CONSIDERATIONS

The process of making hardware functional has three major stages: design, manufacturing, and testing. One specification can be written to include all of these stages, or separate specifications can be written for each. Specifications that include all stages are often referred to as "stand-alone" specifications.

The stages can be further subdivided. Manufacturing specifications can be divided into materials and fabrication (or processing) subtypes. Testing specifications can be divided into prototype (or qualification) and acceptance (or production) testing subtypes. Prototype or qualification tests are comprehensive tests performed on first-of-a-kind items to prove their capability for meeting all design requirements. Acceptance or production tests are the minimal tests performed on each duplicate to demonstrate that it has been properly assembled and that it will meet specified design performance limits that are affected by the range of manufacturing tolerances.

21

A wide range of specification types may be defined. In the end, all must contribute to the same goal and must completely cover the three major stages. The distinctions among them are not always clear-cut, and this can be a problem in a large company that uses a computerized system to keep track of its specifications. Computerized systems require clear-cut distinctions, logical (natural) or arbitrary, regardless of the types defined.

Some features will be common to all well-written specifications, regardless of the differences among the many types in use. This book covers in detail the most widely used general types. Once a writer has gained an understanding of these, this knowledge may be applied to any type encountered.

4.2 COMPONENT AND SYSTEM SPECIFICATIONS

A *component* is a single piece of equipment that will accept a specified set of input conditions and produce another specified set of output conditions. A *system* is a group of interconnected components that does a greater variety of things; it sometimes could be considered a more complex component. These definitions are variants of those given in the glossary and may help to clarify the oftentimes arbitrary distinction.

From a practical standpoint, divisions into components and systems often depend upon who wants to build what. There are some obvious demarcations. You can buy a predesigned electric motor and pump set, or you can buy the motor and pump separately. However, you would not normally try to buy the motor field coils separately from the rotor, or the pump casing separately from the impeller.

From such practical considerations, then, the specification writer must decide, based upon what potential equipment suppliers normally build, whether he or she will be writing a specification for a system or for several components. A system specification will emphasize performance requirements as opposed to design features. A component specification will set out predetermined performance requirements with which a particular segment of industry is familiar and will place more emphasis on desired design features.

4.3 DESIGN AND PURCHASE SPECIFICATIONS

A design specification calls for the production of paperwork, or software, only. The contract may call for a conceptual, preliminary, or final design. That is, it may call for only enough detail to show how the required functions of the equipment will be accomplished; it may call for enough additional details to define operational, maintenance, and safety requirements, and also provide the basis for a close cost estimate; or it may call for a completely detailed design, with all drawings and parts lists, suitable for sending to a shop for construction of a prototype.

Design specifications are primarily functional. The specify *what* the equipment will do, without saying *how* it will be done. In practice this is often an unattainable goal. The group for whom the specification is being written may want certain previously tried design features included or excluded; this may also be true for certain materials. Completely functional design specifications are not very common.

A purchase specification calls for the production of hardware for delivery. It is usually written around tried and proven designs for identical or similar equipment. Most purchase specifications are for standard components, or modifications thereof, intended for use in more complex systems. Purchase specifications, therefore, cover all stages of design, manufacturing, and testing. They usually go well beyond functionality. They often specify what the equipment will do, and how, when, and under what conditions it will do it. Since the bid price will be based primarily upon the purchase specification, this document must be comprehensive, detailed, and complete.

Procuring hardware by competitive bids for fixed-price contracts is the most common method. Many organizations fear that if suppliers are not told both what to do and how, the lowest bids will be for minimally adequate hardware. This is the main reason why so many purchase specifications go beyond functionality in their design requirements. If, instead, an organization negotiates a cost-plus-fixed-fee contract with a well-known and experienced supplier, they may be more willing to include a strictly functional design section in their specification.

4.4 PRIMARY AND SUPPORTING SPECIFICATIONS

Often, a company will buy quantities of the same or similar equipment for different projects over a period of time. Writing a series of complete, stand-alone specifications will result in a large amount of duplication. The obvious way to reduce this duplication is to put the common requirements into separate, special specifications, which can be referenced for their application to each project. The specification that does the referencing is called the primary or base specification. The referenced specifications are called supporting specifications. Usually, the specification containing the design requirements is the primary specification; it will then reference all necessary materials, fabrication, or testing specifications.

Similar pieces of equipment often differ in size or capacity, range of operation, or other features that can be specified numerically. In these instances the numerical values may be left out of the specification and supplied in a separate data sheet. Such a "nonspecific" document is called a general or generic specification and may be used with any number of data sheets to buy a large number of similar items.

There are two distinct approaches to dividing up material between generic specifications and data sheets. The first is to place only the variable numerical information in the data sheets and leave constant values in the generic specification, making that document as complete (and as restricted in application) as possible. The second approach is to place all numerical information in the data sheets, even that which does not vary, making the generic specification as broad in its range of application as possible. The first approach results in the shortest data sheets with minimal repetition; the second approach may result in considerable repetition in the data sheets, where numerical values are constant. Perhaps the commonest example would be fixed ranges of operating environmental conditions such as temperature, pressure, humidity, and precipitation (for outdoor equipment). When application of a generic specification and its data sheets is limited to one specific project, the first approach is usually preferred. For open-ended application to an undetermined number of projects, the second approach may be the better choice.

4.5 THE SELECTION OF SPECIFICATION TYPES

Any large, complex project will start with the development of one or more systems specifications or their equivalent. If the project is to proceed through the hardware stage, the systems specifications will provide the basis for a number of component specifications. If a component is relatively new, the engineering staff may want to investigate more than one design approach. In this case, several component design specifications may be written. When one approach has been selected, the appropriate purchase specification will be written. If the engineering staff has conducted its own design studies and selected the most promising approach, a combination design and purchase specification will be written. If a component is standard, the purchase specification will be written directly from the system specification.

For purchase specifications, the specification writer must decide between single, stand-alone specifications and primary-plus-supporting specifications. If the writer is working on a one-time-only project, the stand-alone specification is generally better. If the writer anticipates multiple project application for his or her specifications (that is, if they will be used for a number of follow-on projects), the primary-plus-supporting specifications may be better. The specification writer may not have a choice, however. The selection may be dictated by customer requirements, current management policy, or past practice.

No matter how the specification-writing effort is divided up, the final result must be complete coverage of the design, manufacturing, and testing stages with all defined project requirements included. If this is not done, trouble is sure to follow. At best, poor or incomplete coverage will lead to some potentially costly contract renegotiation. At worst, lawsuits could result.

4.6 SPECIFICATIONS CONSIDERED IN THIS BOOK

The titles "Purchase Specification," "Equipment Specification," "Component Purchase Specification," and "Component Specification" are often used interchangeably. "Component Specification" will be used hereafter to designate the purchasing document. For some other type of component document, this title will be modified accordingly.

The stand-alone component specification is discussed in detail in this book. This is the most prevalent type and the source of some of the worst problems. A discussion of component design and test specifications will follow. Component manufacturing specifications will not be considered, since they are infrequently used in industry and, when they are used, they are usually internal documents intended for shop planning. They consist mainly of sequentially ordered lists of process specifications, procedures, and special equipment required for intermediate inspections and tests. They may go so far as to include identification of jigs, fixtures, and special-purpose tooling such as machine tools, curing ovens, or vacuum chambers. They are most frequently written by manufacturing specialists. Finally, system design and test specifications will be discussed, covering a type suitable for large-scale, complex systems that would require a number of component design and/or purchase specifications.

Topic outlines for the specifications covered are contained in the appendixes.

Chapter 5

Specification Precedence and Referencing

5.1 ORDERS OF PRECEDENCE

If an organization sends out primary and supporting specifications or a generic specification with data sheets, an order of precedence is necessary so that the highest-level document may be used to modify the contents of one or more lower-level documents.

For primary and supporting specifications, the one containing design requirements is made primary, since the design of equipment occurs first and determines its functions. This then references any other required specifications for manufacturing processes, testing, packing and shipping, quality assurance, and so on. If any of the requirements of these supporting documents are to be modified, the changes are called out in appropriate sections of the primary specification.

If a generic specification and data sheets are sent out, the data sheets are the higher-order documents. There may be three documents in a package (with the package then comprising the complete specification): a data sheet, a type specification, and a general requirements specification. The rule is that the more restricted document has the higher order. When the requirements of a generally applicable specification must be modified for one particular item, and the modifications are placed in the data sheet for that item, the more general documents may be left unchanged for other applications. The same approach may be used for modifying a group of items of the same type by placing the changes in the type specification.

Figure 5.1 shows the document hierarchy when data sheets, type specifications, and general specifications are used to make up complete specifications. Data sheets define individual items, each of which is identified by its own drawing number or part number. Type specifications apply to a family or group of items that perform the same function and have similar physical characteristics. General requirements specifications define the technology and the codes and standards applicable to a large class of items with similar functions and structural requirements. As the controlling document, a data sheet can modify or limit the requirements of its referenced type specification. Similarly, a type specification can modify or limit the requirements of its referenced general requirements specification as applied to a particular group of items.

At the bottom of Figure 5.1 several kinds of general requirements specifications are shown. As the titles indicate, the scopes of these

Figure 5.1   The multipart specification hierarchy

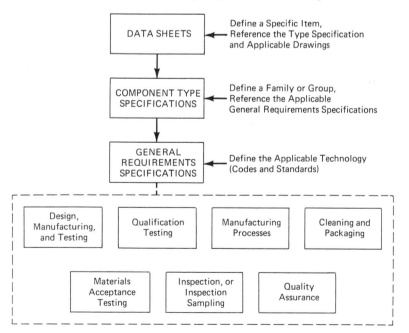

documents may vary considerably. In order to obtain complete coverage, anything not included in a general requirements specification must be included in the associated type specification. Consequently, the relative sizes of these documents may also vary considerably. A type specification may reference several general requirements specifications when the latter are narrow in scope. In any particular instance the choice made usually depends upon the customer, the application of equipment, the branch of industry involved, and the contractor preparing the specification.

5.2 REFERENCING

Documents used to procure the same item of equipment reference each other by two methods: bidirectional referencing and unidirectional referencing. In bidirectional referencing, both the higher- and the lower-order documents specify each other by number and title. In unidirectional referencing, the higher-order document does this, and the lower-order document specifies the higher by title or type name only. When a document is called out by number and title, it is placed in the reference list; otherwise it is not.

The bidirectional method is used almost universally in drawing systems. Usually, there are part or component drawings, subassembly drawings, major assembly drawings, and a top assembly drawing. Higher-level assembly drawings call out all lower-level subassemblies and parts. Each lower-level drawing—except those for standard hardware such as bolts, nuts, and washers—calls out the next higher assembly.

Multiple specification systems have a less rigid structure; here, the unidirectional method is preferred. Changing to a different supporting or base specification is facilitated by the fact that the change must be made in only one document. This method also avoids long reference lists in the lowest-level specifications.

The choice between these two methods is determined largely by the fundamental difference between what is presented by specifications and by drawings. Specifications present requirements (what to do); drawings present the satisfaction of those requirements (how it is done). A single change in a fundamental requirement in one specification frequently brings about changes in several drawings. The more rigid bidirectional referencing method for drawing helps ensure that all necessary changes are made.

The lowest-level specification should reference all of the applicable codes and standards; it should be the base technology document.

Higher-level specifications or data sheets should, if possible, reference only other specifications issued by the specification writer's organization. Often, this rule is difficult to apply. At the very least, it should be followed to whatever degree is possible and suitable. One common instance in which the rule cannot be followed is when the higher-level document must change a code or standard used by the lower-level document.

5.3 SPECIFICATIONS WITH DRAWINGS

When specifications are sent out with drawings, the order of precedence is not as simple. As Figure 3.4 implies, a number of drawing types have a relationship to specifications and their requirements. A discussion of these drawings is contained in appendix A, which also gives the orders of precedence when necessary. Referencing between specifications and drawings is handled in the same way as it is between specifications. Final manufacturing drawings, as shown in Figure 3.4, are not discussed. These drawings are meant only for internal use by the shop; indeed, these drawings are often considered proprietary when none of the others are.

At this point, consideration of the division of requirements between a specification and its accompanying drawing is in order. Requirements that could go in the specification are frequently placed in drawing notes, and if practice is not consistent or controlled this can lead to confusion and oversight. The most common general rule for division of requirements is to make the drawing a design document only, placing manufacturing, testing, and other requirements in the specification. This rule should require no further discussion. Another often-used rule is to limit the drawing to geometric (or arrangement) and dimensional (or sizing) requirements, to the maximum extent possible, and to place other design requirements in the design section of the specification. This rule is subject to flexible application. When national standards primarily concerned with geometry and dimensions are imposed, such as for flanges and electrical connections, these may be called out in the drawing notes. When general-purpose parts and components built to accepted industrial standards are specified, these may also be called out in the drawing notes or in a parts and materials list on the drawing. Codes and standards concerned with design methods are always imposed by the specification.

Chapter 6

Component
Specification
Structure

6.1 STRUCTURAL LEVELS

The division of component specifications into subject matter at the first levels will be the starting point for this discussion, with the three common first-level structures followed by their second-level structures. This will provide the opportunity for a more detailed treatment of each section.

In any organization, the first-level specification structure is usually fixed, whether or not it corresponds to one in this book. The second-level structure presented here is recommended. It fits within the associated first-level structure, and it can readily be adapted to some other first-level arrangement. The third-level structure detailed herein is suggested only. Its primary function is to show what would be contained in a very thorough specification. In many instances a number of these points could be omitted. In any instance the writer must be familiar with the equipment covered, must know the requirements for its particular application, and then must use good judgment in arranging the material.

Whenever possible, a chronological order is followed so that design precedes manufacturing and manufacturing precedes testing. Such orders are natural, easy to understand and use, and facilitate dividing up the work among various groups working to the specification. Most logical orders are based upon opinions concerning the relative importance of topics. Such opinions, usually based upon a few particular applications, are always open to criticism. Also, they never seem to work as well for applications other than those originally considered. Heavy machinery, light mechanisms, electromechanical devices, electrical equipment, electronic devices and all their various combinations can serve as bases for defining a number of arguable logical orders.

For identifying sections, subsections and paragraphs, the decimal numbering system has become almost universal. Alphabetical or alphanumeric designations are used only for appendixes.

6.2 COMPONENT SPECIFICATION FIRST-LEVEL STRUCTURE

6.2.1 The Standard Form

A six-section format for the first level will be referred to as the standard form. These sections are:

1. Scope
2. Applicable Documents
3. Design Requirements
4. Manufacturing Requirements
5. Testing Requirements
6. Other Requirements

6.2.2 The Modified Standard Form

An eight-section modification of the standard-form first-level format is preferred by several government agencies. These sections are:

1. Scope
2. Applicable Documents
3. Design Requirements
4. Materials Requirements
5. Fabrication Requirements
6. Testing Requirements
7. Packing and Shipping Requirements
8. Other Requirements

6.2.3 The Military Form

A six-section format preferred by the military, and hence used by a large segment of the commercial industries, is as follows:

1. Scope
2. Applicable Documents
3. Requirements
4. Quality Assurance
5. Preparation for Delivery
6. Ordering Data

6.2.4 Additions to the First Level

The standard and modified standard forms are sometimes expanded to seven or eight and nine or ten sections, respectively, by adding one or two more sections for Quality Assurance Requirements and Ordering Data.

Whenever the failure of a component could lead to an expensive catastrophe (as in the aerospace industry) or a threat to public health and safety (as in the nuclear energy industry), quality assurance activities assume an extremely significant role. Requirements for quality assurance during the design, manufacturing, and testing stages may be built into the respective specification sections for those stages. However, many engineering organizations prefer to place these requirements in a first-level section of their own. This facilitates their review and monitoring by specially trained personnel. Other engineering organizations think that any extensive quality assurance requirements should be placed in a separate document of their own because quality assurance is not directly related to getting the equipment built, tested, shipped, installed, and operating.

Generic specifications and data sheets were mentioned briefly earlier. If a specification is to be given wide distribution, accompanied by a large number of data sheets, these data sheets can be lost or misplaced. Some companies prefer to place the data sheet material, as tabulated ordering data, in a separate first-level section at the end of the generic specification. The ordering data section may then be modified as required for application of the specification to similar projects.

6.3 SECOND-LEVEL STRUCTURAL LOGIC

The first-level formats for the standard and modified standard forms show the chronological ordering of design, manufacturing, and testing. Second-level formats should also show a chronological ordering based

32

on how things are done, items are used, or actions are taken. If this is not always possible, the ordering should be based on some logical or arbitrary structure imposed by the specification writer or the organization.

Appendix B contains a detailed topic outline of the second-level format of a standard form component specification. You may wish to refer to this outline as you proceed through the subsequent sections of this chapter, which discuss the format in detail. This outline is also meant to serve as a guide for specification writers as they work. For that reason it is as comprehensive as possible. Many items may be omitted for a particular specification, and a different ordering of the items may be better in places. The good judgment of the specification writer must determine the final form of the work.

By proceeding through the logic of the second-level formats for each of the main sections, you will see how the process works and will be able to establish an alternative logic if one is deemed advisable for your specific area of engineering. The author's experience has been mainly in the field of mechanical engineering, and the logic presented may show a definite bias in that direction, in spite of efforts to make it as general and unbiased as possible.

The logical organization of the scope section is quite simple. First, give a brief description of the equipment covered by the specification. Then, describe the limits for applying the specification to items of equipment of that type. If the specification covers only certain sizes or operating ranges, indicate this. Next, describe the extent of work covered by the specification. Indicate whether certain phases of design, manufacturing, testing, and so on, will be performed by someone other than the supplier. Finally, if the specification writer followed certain codes or standards in preparing the specification, these documents may be identified, placing the specification in perspective.

The logical organization of the applicable documents section is also very simple. It proceeds from the documents "closest to home" to the most generally applicable documents—that is, from other documents originating in the specification writer's organization to the nationally used codes and standards. All other documents are placed in between. This section also includes whatever standard statements (also known as "boilerplate") are used by the specification writer's organization to govern the extent of applicability and handling of conflicts between documents.

The imposition of requirements begins in the third main section of the standard form specification. This section—design requirements—is only partially chronological. It is a logically organized section, based on the following considerations. Every piece of equipment has a function or functions to perform, more or less complex, and ways to perform them. The functions, or operations, are expressed as parameters having ranges, limits, or fixed values. The equipment will be designed to meet these requirements, and verification tests may eventually be required. Operational design (or performance) criteria are the first group of items to be defined under design requirements.

Every piece of equipment reacts in some way to its own operation or performance. It may be said that by its operations it creates an environment for itself which it must be capable of handling or surviving. Its ability to do this is determined by structural requirements that go beyond the merely geometric or size requirements necessary for meeting operational requirements. Structural design (or reaction limit) criteria are the second group of items to be defined under design requirements.

Without too much effort a designer can find several ways of designing a piece of equipment to accomplish a specified operation. For a large number of similar operations or functions there are standard workable designs of components. All have their advantages and disadvantages. Based on past experience, the specification writer or the organization may wish to require certain designs, or at least limit the selections available to the designer. Design features are the third group of items specified in this section.

Closely related to the actual design and selection of specific design features is the selection of materials. Indeed, many aspects of the final design will depend upon the physical properties of the materials selected. When all equipment requirements are taken into consideration, design and materials are difficult to separate. Materials selection is the fourth group of items specified in this section.

After the design has been finalized and completed, the only task remaining is to record it in drawings, calculations, and reports. Design documentation is the fifth and final group of items specified in this section.

The organization of the fourth main section of the standard form specification—manufacturing requirements—is straightforward and chronological. First come the items governing materials procurement. Second come those governing manufacturing and fabrication processes (if this distinction is made). Items for marking and identifying the finished product come third, and those for cleaning it prior to testing are fourth. A separate documentation subsection has not been included here. Manufacturing, more than anything else, is subject to numerous quality assurance requirements, including associated documentation. The requirements for manufacturing documentation are assumed to be woven into the preceding subsections through the use of appropriate codes and standards or a referenced quality assurance specification.

The organization of the fifth main section is chronological, based on the fact that prototype or qualification testing is accomplished prior to acceptance testing of production units. The items are ordered as those for prototype testing, acceptance testing, and test documentation.

The organization of the sixth and final section may be recognized as chronological, including items for preparation for shipment, installation documentation, and operational documentation.

6.4 SECOND-LEVEL STRUCTURE—STANDARD FORM

6.4.1 Scope

The term "scope" has almost entirely supplanted the older term "introduction" as the title for the first major section of an engineering specification.

The first item is a brief, general description of the equipment covered by the specification. For standard equipment this may be just its name, perhaps preceded by a widely used type designation. If the equipment is a component in a system, the function of the system may be included.

The second item gives any limitations on the application of the specification to all such pieces of equipment described in the first item. It may restrict application based on size, weight, capacity, classification, standard functional limits, and so on. This item, together with the first, should clearly indicate what equipment is covered and what is not.

The third item gives the extent of work covered. Does the specification call for design only? Design, manufacturing, and delivery for testing? Or design, manufacturing, and testing? Will the supplier just deliver the equipment or also provide installation and checkout services? All such questions should be answered by this item. For more clarity, some organizations require this item to describe both the work covered and the work not covered or done by others.

The fourth item identifies any codes and standards the writer used for preparing the specification. For some equipment in certain applications, conformance of specifications to approved national codes and standards is required. This requirement may be imposed by the final equipment user or by an agency of the government. Including this information places the scope of work in perspective for the equipment supplier. These documents are not listed in the applicable documents section. This section is purely informational and should not be used to impose any requirements on the supplier. All specifications include scope sections.

6.4.2 Applicable Documents

The applicable documents or references section is also common to all specifications and lists all additional documents required to make a complete specification. These documents may be divided into two main groups: those that are sent along with the specification to all prospective bidders, and those nationally used codes and standards, available to everyone, which the bidders will be expected to supply for themselves. The former group may be divided into two subgroups: supporting specifications that are written and controlled by the same organization writing the primary or base specification, and specialized documents that are not generally available but are required by the nature of the project or equipment.

The first subsection under applicable documents, numbered 2.1, is usually a statement concerning the documents as a whole. The following are four standard statements:

a. The following documents, of the revisions specified in the contract (or purchase order), apply to the extent specified herein.
b. The following documents, of the indicated revisions, apply to the extent specified herein.
c. The following documents, of the revisions in effect on the date of invitation to bid, apply to the extent specified herein.
d. The following documents, of the revisions in effect on the date of the contract, apply to the extent specified herein.

The first of these statements is most often used when several documents are included that were written by the organization originating the specification. The specification writer probably has little or no control over these, and some may be undergoing revision while the writer is producing his or her document. The most efficient practice is to identify these revisions in the last document prepared. The second statement requires the specification writer to determine which revisions will be used. If soliciting bids takes long enough, a last-minute updating of document revisions may be required. The third statement effectively sets up current documentation as the basis for bid preparation. The fourth presumes that revisions occurring between the date of invitation to bid and the date of signing the contract will not change anything significantly; otherwise, some renegotiation may be necessary. Some nationally used codes and standards are revised on scheduled dates, and the contents of the revisions are made known beforehand; however, an unexpected major revision can cause problems.

Most of the time, only one standard statement subsection is used, and it is the first item under applicable documents. However, if a number of specialized supporting specifications are included, two may be used. The first would be similar to the first or second statement above for these documents. The other would be like the third or fourth statement for the codes and standards, and it would be placed immediately preceding the codes and standards, with appropriate renumbering of subsections afterward. The wording of each subsection should identify the listings to which it is applicable.

A less common but perhaps more logical practice under the above circumstances is to place the standard statements at the third level, immediately following the second-level group headings. In outline, then, such an applicable documents section could have the following structure:

2. Applicable Documents
   2.1 Company Documents
      2.1.1 Standard statements concerning applicability
      2.1.2 Specifications list
      2.1.3 Drawings list
   2.2 Codes and Standards
      2.2.1 Standard statements concerning applicability
      2.2.2 Organization A codes or standards
      2.2.3 Organization B codes or standards
      2.2.4 Organization C codes or standards

This arrangement is less common and will not be discussed further, but the considerations discussed below for the more common arrangement apply equally well here, and a writer should have no difficulty making the necessary adaptation.

All of the statements say that the documents apply "to the extent specified herein." No reference should be listed that is not specifically called out in at least one place in the body of the requirements. This is because the applicable documents section is just a listing—primarily for convenience—and, like the scope section, it is not used to impose requirements. The total of the callouts may apply to the reference in its entirety, or to whatever extent is desired.

A second standard statement is frequently included with the first: "The order of precedence is as shown." If an order of precedence can be assigned, this may reduce the amount of work required to solve conflicts between references.

The second subsection under applicable documents, numbered 2.2, is where the listing itself begins. Listed here are the supporting specifications or documents originating in the specification writer's own company. Each item listed here and in all following subsections should contain the following information, in the order shown:

a. The document identification number
b. The complete document title
c. The revision number or date (if not specified elsewhere)

All of these documents must be supplied to the potential bidders along with the specification.

The order of listing may follow one of two contrasting rules. The first rule is that the order should be numerical or alpha numerical, whichever is appropriate. The second rule is that the listing should follow the order of callouts within the requirements sections. Either approach is acceptable, provided it is followed consistently.

The third subsection under applicable documents, numbered 2.3, lists those "special" documents that are not readily available and that originate outside of the specification writer's company. These are usually compilations of materials test data or generalized design data put together by specialized organizations, such as materials testing laboratories, consulting firms, government laboratories, or research foundations. For practical purposes, their revisions are unscheduled and erratic.

At this point, a warning is in order. Many companies frown upon or even forbid calling out references that they must supply and over which they do not have revision control. It may be difficult to get copies of the correct revision, and this can lead to problems and confusion. The solution is not to reference such documents; instead, copy their applicable contents into the primary specification or one of the supporting specifications. This requires more work on the part of the specification writer, but it may prevent time-consuming problems later on.

The next subsection may be numbered 2.3 or 2.4. It may contain the second set of standard applicability statements, or it will start listing the generally available national codes and standards. If there is an order of precedence, it is paramount here. It is assumed that a company's own specifications take precedence over codes and standards.

Some organizations prefer to separate documents originated by government regulatory agencies from the above. If so, these are placed in a final subsection. This is largely a matter of preference and is significant only when several regulatory agencies have something to say about a particular product.

Some organizations make one exception to the rule that all documents called out in the requirements sections (from section 3 on) be listed in section 2. This exception is for material specifications. The argument is that standard material specification numbers are well known to everyone in the industry. Also, material specifications always apply in their entirety—never to some limited extent determined by the requirements section. Finally, listing every material specification could result in an excessively long applicable documents section.

6.4.3 Design Requirements

The following discussion of the three main sections of the specification will be based on the standard form as shown on page 00. Converting to either of the other forms will be relatively easy.

The first main group of items under design requirements is made up of the operational design criteria. These will determine the functional characteristics of the equipment covered. All applicable design codes, standards, and specifications should be called out first. Usually, these are identified by an abbreviated name or acronym for the originating organization and the document number. Sometimes, they are identified by their list item number from section 2 or by duplicating their full listing from section 2 (without the item number), which results in rather lengthy callouts.

If only a part or parts of a document will be used, these may also be identified here (the "extent specified herein"). Care should be exercised if applicable parts are referenced by section or paragraph numbers only. An imminent revision of the document may change these. A useful alternative is to say something like "The . . . shall be in accordance with . . . as specified below," and to place the section and paragraph identification in appropriate locations along with individual requirements.

Following the initial callout of codes and standards, the normal operating (performance) parameters should be specified. Immediately afterward, there should be a definition of the normal operating conditions (environment) for the equipment, including the expected service life. If applicable, the service life should be broken down into a table giving required total operating times at different operating levels. If maintenance intervals are specified, they should be included here.

Next come the abnormal operating parameters, if such are to be specified. These will be the normal parameters as affected by any adverse operating conditions. Immediately afterward, there should be a definition of the abnormal operating conditions, their time intervals, and the number of times they should be expected to occur during the service life of the equipment.

Specifications follow for any required adjustable and nonadjustable set points and limits, including the adjustment ranges. Finally, there should be a listing of any multiple (redundant) inputs and outputs required for connecting the equipment to its associated system or systems.

The second main group of items under design requirements includes the structural design criteria. The operational design criteria are related to what the equipment does; the structural criteria are related to how the equipment reacts. There are two basic types of equipment reactions: reactions to the external environment in which it is placed, and reactions to the internal conditions it creates by its own operation. These internal conditions are usually referred to as *loadings*.

The first items are, of course, the applicable specifications, codes, and standards. Because of their direct relationship to safety, reliability, and durability, structural design criteria are probably subject to more national codes and standards than any other design requirements. Next come the steady-state loadings, both normal and abnormal (if abnormal ones are postulated), which the equipment must be designed to withstand. The criteria for design evaluation should be included if they are not in the references.

Then come the transient loadings, both normal and abnormal. If there are many of these, or if they are complex in nature, they may be placed in an appendix, which is referenced here. If the criteria for evaluating the effects of these loadings on the design are different from those for steady-state loadings, this must be explained. In some quarters, steady-state or time invariant loads are called *dead loads*, and transient or time-varying loads are called *live loads*.

After this come the expected loading combinations. All steady-state loadings may not occur together. All transient loadings seldom, if ever, occur simultaneously. The superimposition of transient loadings on steady-state loadings is subject to much variation, even in the same system. The specification writer must determine and describe the complete range of loading combinations to be considered for the design.

Finally, there are the requirements for analysis—that is, how the analysis is to be accomplished. Are there certain formulas or widely available computer codes that must be used? Are specific scale factors, safety factors, or correlation coefficients to be used? Any specific limits on the methodology should be set down at this point.

The third main group of items includes the required design features. Basically, the design criteria give the "what" requirements, and the design features give the "how" requirements. These may be divided into mechanical, structural, and electrical features (details), or some other grouping suitable for the equipment. Each division should include references to any applicable codes, standards, or specifications.

The fourth main group of items includes the requirements for materials selection. These requirements are imposed if the specification writer is not willing to give the manufacturer complete freedom to select the materials. If the equipment specified is very complex, it may be helpful to divide its parts into groups. The following are general, frequently used groupings. For any particular piece of equipment, however, the specification writer may devise another more suitable scheme.

1. Parts may be grouped on the basis of criticalness and safety—those that are critical to operation and safety, those critical to operation only, and those that are noncritical—that is, those whose failure would necessitate shutdown and create a hazard for the operators and perhaps the public, those whose failure would necessitate shutdown without endangering anyone, and those whose failure would not require immediate shutdown.

2. Parts may be grouped on the basis of their material type, such as metallic parts, nonmetallic parts, and composite material parts. This method of grouping is readily subject to expansion.

3. Parts may be grouped on the basis of their functionality, such as stationary parts, moving parts (rotating or reciprocating), shafts, seals, fasteners, mechanical parts, electrical parts, lubricants, and so on. This method of grouping is very common and offers a wide range of possibilities.

4. Parts may be grouped on the basis of their durability, such as lifetime parts, those that should be replaced on a scheduled basis, and those that have to be replaced as required. This is probably the least common method of grouping.

5. Parts may also be grouped on the basis of a suitable combination of any of the above methods.

Whatever approach is used, the specification writer is well advised to consider common practice in the segment of industry supplying the type of equipment specified.

After the parts have been grouped, the next step is to specify the material or materials for each group or for the individual parts within a group. Broadly speaking, all commonly used materials may be divided into four categories: required materials, preferred materials, allowable (alternative) materials, and disallowed materials. Failure to use required materials or the use of disallowed materials will result in bid rejection. Preferred materials are worth more than allowable materials in the evaluation of acceptable bids.

Of course, the specification writer may not wish to use all of the above categories. He or she may not know all of the commonly used materials. Instead, the writer may use two or three of the above categories and include a statement such as, "The use of other materials is subject to purchaser approval."

The fifth main group of items includes the requirements for design documentation. These are the design reports and drawings the specification writer (and his or her organization) will require from the equipment supplier. Reports may cover operating and structural characteristics of the equipment. Along with the drawings, the reports should show how the proposed equipment meets the design requirements of the specification. The level of detail required should be indicated. The suppliers should know if you want only results from their analyses, or enough detail to permit checking of their calculations. Remember, more detail increases both the time required to prepare a report and its cost.

Required drawings usually fall into three classes: proposal drawings, initial (outline and connection details) drawings, and final assembly (cross-sectional) drawings. The last are the drawings that eventually accompany operation and maintenance manuals. A materials list should accompany or be part of the final assembly drawings.

Requirements for proposal drawings are necessary if the specification writer's organization has to produce interface control or installation drawings based on them soon after the winning bidder is selected. Otherwise, the initial drawings will be used for this purpose. The specification must indicate when reports and drawings (other than proposal drawings) are due. Usually, this is given as a fixed time period after receipt of the contract.

The specification must also indicate for each document whether it is to be submitted for information, for review only, or for review and approval. If a document is submitted for information, the supplier expects to hear no more about it. If a document is submitted for review, the supplier may expect some questions; if any mistakes are found, the supplier will expect to correct them. If a document is submitted for review and approval, certain work will stop until formal approval has been transmitted. The suppliers may plan some revision prior to final approval and will provide for this in their estimates. Therefore, requiring both review and approval of a document will increase equipment cost.

6.4.4 Manufacturing Requirements

The next main section of the specification contains manufacturing requirements. The first main group of requirements in this section is for material procurement. Initial items are applicable specifications, codes, and standards that will govern material procurement. Then come requirements for tests of basic material properties of each lot or quantity of material prepared for the order. These tests are usually destructive in nature and are performed on representative samples of the materials. The specification should indicate whether archive samples are to be stored or delivered, for quality assurance purposes.

Next come requirements for nondestructive examination of materials and associated acceptance criteria for the methods used. These examinations are performed on materials that go into the equipment, and the most common are for specialized castings and forgings.

Alternatively, the requirements imposed here may be for testing representative samples of simple finished parts, taken from large quantities procured for use in more complex assembled products. In this situation, tests may be either nondestructive or destructive. If the testing is thorough, checking limits as well as design function, it is usually destructive. An example would be testing samples of electronic parts, which will be subjected to severe environments during use.

Finally, there are requirements for documentation of the above tests and examinations. Documentation usually takes the form of certified materials, test reports, or certificates of conformance. The complete test reports require more work on the part of the supplier and will increase the total cost of the equipment. The less costly certificates of conformance merely state that a supplier has met all requirements, without providing actual data or evidence. Normally, the complete reports are called for only if they are necessary to meet the requirements of applicable national codes or standards.

The second group of requirements in this section is for fabrication or manufacturing processes. Once again, the initial items are the applicable specifications, codes, and standards. Next, the allowable processes, modifications to these processes, or exceptions to standard variations in these processes are listed. Closely related items are limitations on applications of allowable processes.

Processes here involve the application of thermal, chemical, or electrical energy. These are processes such as heat treating, welding, plating, surface treatment, and so on. Such processes require careful control, often by sophisticated means. Minor deviations in control can cause a significant deviation in the final results, and exacting specifications are often necessary.

The next items are, quite logically, the disallowed processes. These are usually based on previous unsatisfactory experiences. Disallowed processes are frequently the source of seemingly endless debate among experts, and the main source of exceptions from potential suppliers.

Following the requirements placed upon processes are those placed upon the examination of finished parts. These requirements must include the necessary acceptance criteria. As for materials procurement, these are nondestructive examination requirements. One set of requirements applies to the so-called raw materials, and the other to the processed parts that are ready for assembly.

The final items are any special dimensional measurements that must be made and recorded beyond what are required by the piece part drawings. Usually, these are related to some specific design calculations.

The third main group of requirements in this section includes the requirements for identification. Items of information that must appear on the nameplate are usually first, but the order in this group is immaterial. The bare minimum requirements are for the drawing or part number and the serial number. Other requirements for information may be determined by the specification writer's organization.

Next, list any requirements for warning tags to be placed on the equipment for precautions or critical operations during installation, startup, and checkout. Finally, list any other markings, such as movement directions, inlets and outlets, connection identification, and so on.

The fourth main group of requirements in this section includes the requirements for pretest cleaning. The only items here are the required or allowable cleaning materials and methods. Cleaning material residues are often a major concern, and the supplier is required to determine the materials, methods, and associated procedures and submit them to the purchaser prior to use. The submittal may be for review only, or for review and approval.

6.4.5 Testing Requirements

The last of the three main sections of the specification is for testing requirements. This section is divided into prototype (qualification) testing, acceptance (production) testing, and test documentation.

Prototype testing requirements come first. The initial items, as usual, are the applicable specifications, codes, and standards. Next come the requirements for structural testing, which will verify that the equipment will meet the structural design criteria. This is particularly important when safety considerations are involved.

Immediately following are the requirements for operations or performance testing, which will verify that the equipment will meet the operational design criteria. Often, these tests are not required when proper control of part dimensions is sufficient to guarantee proper operation. This is frequently the case with equipment designed to handle flowing fluids, if flowpath geometries with well-established characteristics are used.

After that come any requirements for environmental or life tests. These are subdivided into normal environment testing and abnormal environment testing. These tests will verify that the equipment will meet its service life and maintenance interval requirements under the specified conditions. This is particularly important when equipment replacement would be very difficult.

Finally come any requirements imposed on delivery or disposition of the test unit. Will it be scrapped? Will it be stored by the supplier? Will it be shipped to the purchaser for storage? Will it be refurbished and delivered for use? If it is to be delivered for use, a requirement should be included for documentation of the rebuilding, including a list of items replaced or reworked and an inspection report.

The second main group of requirements includes those for acceptance testing. The above prototype tests are usually performed on one unit only, while acceptance tests are performed on every unit delivered. Once again, the initial items are the applicable specifications, codes, and standards. To ensure uniformity and safety, acceptance testing of many types of equipment is covered by industrial standards.

The next items cover assembly verification tests. If the equipment contains moving parts, their operation must be checked to ensure that the equipment has been correctly assembled and adjusted.

After that come the requirements for proof tests. These cover structural or electrical tests in which some loading is set at a maximum condition which the equipment must survive without damage. Such tests verify the safety requirements imposed upon the equipment.

The final items are the requirements for operational or performance testing, usually a simplified version of the prototype operational tests. For many types of equipment there are no operational (performance) tests. As before, equipment for handling flowing fluids is a good example. If the production unit geometry duplicates the prototype geometry (within the limits of manufacturing tolerances), it is assumed that it will perform properly.

The third main group of requirements includes those for test documentation. Requirements for the documentation of prototype tests, of course, are usually much more extensive than those for acceptance tests. Prototype test documentation requirements will be discussed, with indications of how to simplify them for acceptance testing.

The first items under this group are the requirements for a test plan or plans to be submitted to the purchaser for review, or for review and approval. The degree to which the purchaser will want to review and approve these plans depends, to some extent, upon how much of the specified testing is done in accordance with national codes and standards. Next come the requirements for instrumentation, for the equipment tested and for the test facility. The latter presumes that there are requirements beyond what would normally be expected. Then come the requirements for drawings of the test setup, special test fixtures, and so on. After that come the requirements for written test procedures covering pretest checks, calibrations, precautionary measures, details of the test run or runs, and so on.

The final items are requirements concerning the form and organization of test results—the test report format—and any verification documentation to be included. The latter would include calibration certificates and perhaps verification of traceability to the National Bureau of Standards.

For acceptance test documentation, the first items are the individual tests to be performed. Next come the parameters to be measured and recorded. The above may, to some extent, be dictated by national codes and standards. Then come the requirements for an acceptance test procedure. This is usually a single document containing sketches of any special fixtures or setups. After that come requirements for presentation of test results. These may take the form of a test data sheet. Finally, instrument calibration requirements are treated. For acceptance tests, this usually takes the form of a certificate of calibration status rather than individual certificates for each instrument.

6.4.6 Other Requirements

The fourth and final requirements section of a standard-form specification is the section for all other requirements—for the areas of preparation for shipment; installation documentation for the use of those who receive, install, and check out the equipment; and operational documentation for the use of those who will operate and maintain the equipment over its service life. If there are any requirements for special services from the supplier after delivery and at the site of equipment installation, they should also be included here.

The first main group of requirements, then, is for preparation for shipment. As usual, the initial items are any applicable specifications, codes, and standards. Next come the requirements for final cleaning. Here again, the items cover required or allowable cleaning materials and methods. The other comments under pretest cleaning also apply here.

Logically following are any requirements for the use of preservatives. The preservatives may be applied directly to the equipment surfaces or may be a part of the packing. The specification writer should have some idea of how long the equipment may be stored prior to installation, and under what conditions.

The next items are the packing requirements themselves. These are also related to the length of time and conditions of storage, as well as transportation. Presumably, the specification writer, after some study and consultation, has some idea of how such equipment is usually shipped. The writer should consider what sort of shipping damage could occur and how to protect the equipment. This may be left up to the supplier, who is required to provide a packing and shipping plan, submitted for review, or for review and approval.

Finally, there are the requirements for labeling the packages or crates. Some items of information appear on almost all labels, such as the shipper's name, the receiver's name and address, the weight, and the drawing number. The specification writer's organization may require more information, such as the contract number, the invoice number, and the serial number(s). The receiver may want labels on more than one side of the crates, so that close stacking in storage will leave at least one in view.

The second main group of requirements is for installation documentation. The initial item(s) are the requirements for receiving inspection instructions. These usually are for complex, nonstandard pieces of equipment, which could be subject to shipping damage that is not easily detectable. How should the receivers make sure that the equipment is in the same condition as it was when it left the factory?

The next items are the handling instructions. How should the equipment be moved and lifted into place? For small, light pieces of equipment this is no problem, unless, as in the case of optical equipment, the supplier wants to say "Handle with Care." For large pieces of equipment, however, handling may be critical. Many such pieces have special lifting points and attachments.

Then there are the requirements for installation instructions. After the equipment is in place, how is it fastened down? What is connected first and last? How are the connections made if the procedure is not self-evident?

Finally, there are the requirements for checkout procedures. After the equipment has been installed, what must be done to verify its operability as a functional part of the system? Also, what must be done to ensure that the equipment has been properly installed?

The third and last main group of requirements is for operational documentation, mainly for operation and maintenance manuals (which may come as a single document). These manuals may seem to be standard, not needing any requirements, but the specification writer can never be sure. Sometimes, for simple, standard equipment, a statement that the manuals are required is sufficient. Their cost will then be included in the total cost. Also, the specification writer's organization may want one reproducible copy for their own use. This will affect the cost, especially if several assembly drawings are included. The specification writer may want to be sure to include certain things in the manual. For complex pieces of equipment there may be a comprehensive manual that references or includes standard manuals for general-purpose components supplied by subtier contractors. Requirements may be imposed upon the organization of this overall manual.

Finally, there are requirements for a special tools and spare parts list for the maintenance crew. This will be used for initial ordering of spare parts and special tools. Again, the basic purpose of these requirements is to make certain that the supplier provides complete information on what it will take to keep the equipment operating throughout its service life.

Now that this discussion is complete, there is a point worth further consideration. Each of the major requirements sections begins with a callout of the codes and standards being applied. Often, the same callout covers all areas. Although repeating a callout in each major section may seem redundant, it is advisable. This is how "the extent specified herein" is established for each of the areas of design, manufacturing, and testing. If, for instance, one wanted to use a code for design and testing only, this approach facilitates applying some other document in the manufacturing area. Inexperienced writers sometimes incorrectly place a general coverage requirement for a particular code in section 2, in an attempt to avoid multiple callouts. In addition to being incorrect, this makes modifications to or restrictions on applications of the code in specific areas more difficult and possibly confusing.

6.5 SECOND-LEVEL STRUCTURE—MODIFIED STANDARD FORM

Having gone through a detailed consideration of the second-level format of the standard-form specifications, we will now consider how it may be converted to the modified standard form. This section is brief because the two forms are similar.

6.5.1, 6.5.2 The Scope and Applicable Documents

These first two main sections are identical to those of the standard-form specification.

6.5.3 Design Requirements

This section is the same as that for the standard form, except that the requirements for materials selection are omitted. These are placed in the following materials requirements section instead. Most specifications will be written for equipment with components of standard designs. For these designs a relatively small group of materials is known to be satisfactory, and these materials are widely used. Therefore, materials selection is routine once the design features have been determined, and the requirements for it may be placed elsewhere for convenience.

6.5.4 Materials Requirements

In the modified standard form this section contains the requirements for materials selection and procurement; that is, the materials selection requirements discussed under subsection 3.4 of design requirements of the standard form, and the materials procurement requirements discussed under subsection 4.1 of manufacturing requirements of that form. There is some logic to the combination of selection and procurement in the same section. Often, the selection of a material and its related form has some bearing on its procurement requirements. An example is the selection of a casting or a forging for a major structural component.

6.5.5 Fabrication Requirements

The fabrication requirements of the modified standard form are the same as those for the standard form with the material procurement requirements omitted. At this point we should touch upon the fine and somewhat nebulous distinction between manufacturing and fabrication. If a structural component is made by welding several pieces together and then performing machining operations on the weldment, the component is said to be *fabricated*. If a component starts with a casting or a forging, or a piece of bar stock or formed plate, and is then subjected to machining operations, the component is said to be *manufactured*. The distinction arises from the uniqueness of welding processes and their associated heat treatment requirements. For complex components the distinction may be hard to make. Entitling the section "Fabrication Requirements" implies that welding requirements will play a predominant part. As a matter of fact, welding processes are much more often the subject of national codes and standards than are other manufacturing processes.

Please note that a fabrication specification is not the same as a welding specification. A welding specification is concerned with the application of a particular welding method, such as gas tungsten arc welding, to a particular material, such as austenitic stainless steel. A fabrication specification is concerned with all of the welding processes, materials, and heat treatments applicable to the production of a particular component or group of similar components.

6.5.6 Testing Requirements

The testing requirements section of the modified standard-form specification is the same as for the standard form.

6.5.7 Packing and Shipping Requirements

These requirements in the modified standard-form specification are the same as those in the first subsection under other requirements in the standard form. There is some logic in separating them from the final documentation requirements.

6.5.8 Other (Additional) Requirements

These requirements in the modified standard-form specification are the same as the final documentation requirements in the second and third subsections under other requirements in the standard form.

6.6 SECOND-LEVEL STRUCTURE—MILITARY FORM

The second-level structure of the military form for component specifications will be discussed by relating it to the standard-form specification.

6.6.1, 6.6.2 The Scope and Applicable Documents

These first two main sections are identical to those of the standard-form specification.

6.6.3 Requirements

This third main section of a military-form specification contains all of the requirements pertaining to the equipment itself; that is, it contains all of the requirements of sections 3, 4, 5, and 6 of the standard-form specification, with the exception of the preparation for shipment subsection of section 6. The second-level breakdown for this form may be obtained by appropriate renumbering of those standard-form sections. This places all the equipment requirements in one section.

6.6.4 Quality Assurance

The fourth main section of the military form is concerned with quality assurance. In recent years quality assurance activities have become increasingly important and are gaining recognition as a separate discipline.

This section usually starts out with requirements for a general quality assurance program in the supplier's facility or an approved quality assurance manual suitable for the products involved. Following that may be requirements for a particular quality assurance plan, oriented toward the product covered by the specification. Then there may be requirements for a quality assurance document index. This is a cross-reference list that relates the requirements of the plan, including any applicable codes and standards, to the supplier's shop records. Next may come requirements for an inspection and test plan or plans. All of the necessary inspection and test measurements, in their proper sequence, would appear in the plans. After that may come requirements concerning the qualification and certification of processes used and of the personnel using them. Next would come requirements for the handling of items that do not conform to the inspection and test acceptance limits, what may be done in the way of reworking, and how material should be scrapped. At this point full-blown quality assurance programs often concern themselves with requirements for auditing the supplier's work for conformance to all of the above requirements. These audits are often considered to be the most important quality assurance activity. Finally, there are requirements for the submittal of records to the purchaser—records that cover the processing history of every critical part in the equipment, and all tests of subassemblies and the complete item.

6.6.5 Preparation for Delivery

This fifth main section of the military-form specification contains the same requirements as found in the preparation for shipment subsection of section 6 of the standard-form specification. The military frequently stores equipment for much longer periods of time under a much wider variety of conditions than any other organization. For that reason this section may be relatively long and may contain a variety of conditional requirements. That is why these requirements are placed in a main section of their own.

6.6.6 Ordering Data

In chapter 1, the use of data sheets for listing quantifiable requirements was discussed. Different data sheets then specified similar items of equipment of different sizes, ranges, and so on. The sixth main section of a military-form specification is meant to contain a sample data sheet for the equipment. The arrangement of data should either reflect the arrangement of related requirements in the third section, or be a more efficient logical arrangement. Instead of data, this sample data sheet contains the paragraph numbers of the related requirements.

The main advantage of this approach is that it enables different groups to use the same specification correctly and consistently for different procurements. Such repeated procurements are most common in the military and other government agencies.

Nonmilitary organizations using the military form often put all of the completed data sheets for a procurement into the ordering data section. This makes a single, complete, convenient package. If they use the same specification for another procurement, they take out the old set of data sheets and replace it with a new one.

6.7 MODIFICATIONS FOR A PRIMARY SPECIFICATION

At this point we will discuss what material is usually placed in separate, supporting specifications, if these are used, and what is left in the primary or base specification. As was stated in chapter 4, the primary specification contains the design requirements. Practically all of the requirements that come under the heading of manufacturing could be placed in a separate specification or specifications. Separate documents could be written for materials procurement and for each major manufacturing process. Since quality assurance is primarily concerned with manufacturing, a separate quality assurance specification could be utilized here.

Qualification testing requirements are frequently placed in a separate specification, with acceptance testing left in the primary document. This is desirable because qualification testing requirements may very easily be left out of the procurement package if the equipment ordered is already qualified. Finally, cleaning, packing, and shipping requirements are frequently placed in supporting specifications. These can vary considerably for the same equipment, depending on where it is to be shipped and how long it will be stored.

In summary, then, a primary specification may do the following in terms of requirements:

1. Contain the design requirements.
2. Reference the material procurement and manufacturing requirements.
3. Reference the quality assurance requirements.
4. Reference the qualification testing requirements, if necessary.

5. Contain the acceptance testing requirements.
6. Reference the cleaning, packing, and shipping requirements.
7. Contain the documentation requirements.

The primary specification will contain all of the section headings of the stand-alone specification, and probably many of the subheadings. Each of these will reference the appropriate supporting specification.

This may seem like a lot of documents to send out for a single item of equipment if you are buying only one design. However, an organization will frequently buy a number of similar types of equipment at one time from the same supplier. A number of primary specifications then can utilize common supporting specifications.

In chapter 5, an arrangement consisting of data sheets, type specifications, and general requirements specifications was discussed. This approach is often used for equipment such as valves and amplifiers, each of which has a variety of types with many things in common. The general requirements specification or specifications are referenced by the type specifications but go far beyond the specialized supporting specifications discussed above. If possible, they list all of the applicable codes and standards, with appropriate referencing, and therefore are the basic technology documents. The type specifications resemble restricted primary specifications. They contain requirements unique to the type covered and reference only the appropriate general requirements specification or specifications. The data sheet defines a specific item and references the related type specification.

6.8 COMPONENT DESIGN SPECIFICATION STRUCTURE

Having completed consideration of the structure of specifications used for purchasing hardware, we now move on to a consideration of specifications used for obtaining finalized designs only—that is, for obtaining equipment designs ready to be sent out to the shop for fabrication of prototype units. Such specifications are frequently written by companies primarily interested in building and delivering hardware, without doing much design work. A six-section design specification will be considered, loosely patterned after the six-section standard-form purchase specification.

6.8.1 Scope

The scope section of a design specification starts out with the name the specifying organization has assigned to the equipment it wants designed. If the name does not clearly indicate what standard or well-known items of equipment it is similar to, this information should immediately follow. Next should come a brief description of the principal function(s) of the equipment. If it will be installed in a system, a description of the segment of the system in which it is placed, together with a description of how it affects system operation, is in order. As before, this section is descriptive only, providing background material, and does not impose any requirements.

6.8.2 Applicable Documents

The applicable documents section of a design specification is similar in structure to that of the standard-form purchase specification. The specifying organization's drawings, if any, will be equipment requirement drawings showing space limitations, connection requirements, and perhaps some internal configuration requirements. As before, this section is purely informational and does not impose any requirements.

6.8.3 Design Criteria

This and the remaining sections, as before, impose the requirements of the specification. The design criteria are the same as those for operational and structural design, subsections 3.1 and 3.2 in the standard-form purchase specification. This material must be complete and comprehensive in a design specification.

6.8.4 Design Features

The section on design features is the same as that covering structural, mechanical, and electrical design features in subsection 3.3 of the standard-form purchase specification. The amount of detail may vary considerably.

In a purely functional design specification this section would not exist. The description in the scope, the complete design criteria, and perhaps a minimal equipment requirement drawing would be sufficient. However, such "pure" design specifications are rare. Every engineering organization writing design specifications has its preferences (or prejudices) for certain ways to accomplish specific functions. They can always find ample justification, based on experience, "expert" advice, or some formal documentation of "recommended practices" for requiring certain features. One of the main concerns, if not *the* main concern, behind excessive specification of design features is cost. The problem is to get an adequate design at a reasonable cost instead of an excessively high-quality design at a high cost or a marginal design at the lowest cost.

Ideally, then, this section should be kept brief, giving the designer the maximum number of options. However, if your organization has some feature it really wants, no matter how small, it should be included. Otherwise, you may end up going through repeated reviews until the designer finally sees it and provides it.

6.8.5 Materials Selection

This section contains the same material covered in the standard-form purchase specification in subsection 3.4. This is another section that could be omitted ideally, giving the designer the widest choice for accomplishing his or her task. Also, this is another area where the specifying organization will most probably have a fund of experiences, opinions, and preferences concerning the best materials to use for familiar parts in numerous applications.

It is hard to say just how comprehensive this section should be. The most probable minimum would be a specification of interfacing materials only. The maximum would be a specification of materials or groups of materials for every type of component expected to be used.

6.8.6 Design Documentation

The final section, covering design documentation, includes the material of subsection 3.5 of the standard-form purchase specification. It is more comprehensive here in the design specification. The coverage of design reports usually would be more thorough and detailed. If a complete package of drawings is required, ready to be sent to the shop, drawing requirements must be more detailed, especially if the specifying organization wants to impose its own structure, standards, and style for drawings, including the use of its own title blocks. A preliminary set of operations and maintenance instructions may be required. This will help to ensure that sufficient thought is given to these matters and a usable design is produced.

6.9 COMPONENT TEST SPECIFICATION STRUCTURE

The design of a new piece of equipment requires that a comprehensive prototype test program be undertaken to verify that the equipment will meet all design requirements. Frequently, manufacturing-oriented companies contract these prototype or qualification test programs out to independent testing laboratories. The document that determines what this program will be is the test specification. Often, test specifications are also written for in-house testing to make sure that the program is understood by everyone involved and documented all in one place. A seven-section test specification, structured chronologically, is discussed below.

6.9.1 Scope

The scope section of a test specification starts out by identifying the equipment to be tested. Then it should indicate all of the general types of tests to be accomplished, such as performance tests, structural tests, environmental tests, and endurance (life) tests. If the equipment is to be tested to failure, this should be stated. If transient as well as steady-state testing is to be accomplished, this should also be stated. Once again, this section is descriptive only and does not impose any requirements.

6.9.2 Applicable Documents

This section will, of course, start off with any standard statements concerning applicability and order of precedence that the writer's organization uses for test specifications. The reference list itself starts off with the design documents. First comes the design specification to which the equipment was designed; this tells *what* it is expected to do. Next comes the design report; this tells *how* the equipment does it. Then the applicable design drawings should be listed. As a minimum, these would consist of the assembly drawing and installation drawing.

Other documents follow, applicable to the testing and supplied by the specification writer's organization. In most cases these would be required when the equipment is being built for some government agency with its own special test requirements that it wants imposed on its equipment.

Finally, the listing includes any applicable national codes and standards. There are standards covering the testing of many common equipment types. There are also standards covering specialized instrumentation for critical and difficult measurements.

6.9.3 Test Objectives
(Test Criteria)

The section on test objectives, or test criteria, should indicate what particular information is to be obtained for each type of test performed;

that is, what information should be obtained, directly or indirectly, for each of the test types of the scope, to determine if the requirements of the design specification have been met. This makes the second-level structure of this section obvious.

6.9.4 Instrumentation

For a complex test, which will involve a number of test runs and where not all measurements are required for each run, this section should start out by indicating what measurements must be made for the evaluation of each test objective (criterion) for its assigned test run. A distinction must be made between instruments measuring equipment parameters and those measuring related test facility parameters. The specification should then state the frequency of measurement (or recording) for each parameter. Shall it be continuous or intermittent, and, if intermittent, how frequent? Should the value merely be monitored and notes made if it varies from the expected level? The above requirements have a significant effect on testing costs, since recording is more expensive than monitoring.

At this point the specification writer may provide an instrument list for the entire series of tests. The list must include all instruments necessary for the above measurements and the range and accuracy required for each instrument. Finally, this section should indicate the requirements for instrument calibration.

For a simple test the instrument list may comprise this entire section.

6.9.5 Set-Up

In addition to the instrumentation, one of the most important components of the test set-up is the test fixture. Usually, the design of the test fixture(s) is left up to the organization doing the testing. However, the organization requesting the testing may have some requirements it wants to impose. These are usually related to expected service installations. Such requirements are imposed by this section.

The design documentation section of the design specification may have included a requirement for initial or recommended operating and maintenance instructions. These should have been included in the applicable documents section above. The organization requesting the testing may have some operational requirements it wants to impose, also related to expected service installations, which will affect the details of the test set-up. These requirements should be imposed by this section.

6.9.6 Procedures

Detailed test procedures and instructions are written by the organization doing the testing. The procedures imposed in this section by the requesting organization are general in nature and are related to the equipment design and the testing required. This section may simply impose a required sequence for the tests described in the scope.

6.9.7 Test Documentation

This section describes all of the records and reports required by the requesting organization. It should cover such things as records of raw data, calibration records or certifications, reduction of raw data to its desired form, and the method of final data presentation. If a formal test report is required, its contents should be specified here.

Chapter 7

System Specification Structure

7.1 SYSTEM TYPES

A *system* is sometimes defined as a group of components connected together in such a way that it can accomplish a number and variety of functions that cannot be accomplished by the components themselves. Usually, a number of the components in a system are adjustable in their functioning. Thus, the number and variety of systems and functions that can be obtained from a given set of components is large. This makes standardization of system specifications through more than one level considerably more difficult than it is for component specifications.

For this discussion all systems may be considered as one of two types. The first is the large-scale system that is put together at the location where it will be used. This type of system uses a wide variety of equipment, which is bought from a number of specialized suppliers, brought to the site, assembled, checked out, and put into operation. An example would be the heat transport or steam supply system of a power plant.

59

The second type is the complex system, with a more limited variety of equipment, which is put together and checked out in the supplier's shop and then shipped to the location where it will be installed and used. This type of system may also use components bought from specialized suppliers. An example would be the electrical and electronic sound system for a theater or sports arena.

7.2 SYSTEM SPECIFICATIONS

System design specifications and system test specifications will be discussed here, along with how they differ from their component specification counterparts. For systems of the first type above, a construction specification, written according to construction specification standards, would be the equivalent of a manufacturing specification. For systems of the second type, an assembly specification or assembly instructions might be written.

The specification structures presented are biased toward the first type but are suitable for the second with some modifications. For small, simple systems of the second type the component specification structures, with suitable modifications, could be used. These modifications will be left up to the writer or organization doing the work.

7.3 SYSTEM DESIGN SPECIFICATIONS

The structure of a system design specification represents a radical departure from that of the component design specification. The first-level structure of the system design specification consists of the following seven sections:

1. Scope
2. Applicable Documents
3. Design Requirements
4. Safety Requirements
5. Operational Requirements
6. Maintainability Requirements
7. Documentation Requirements

A comparison of this structure with that of the component design specification indicates the difference considerations involved. In general, the system design specification emphasizes breadth of coverage, while the component design specification emphasizes depth. The system document addresses areas of requirements largely taken for granted by the component document. The component document must be used to produce a design ready for the shop, while the system document serves as a source of requirements for component design or purchase specifications.

7.4 SYSTEM DESIGN SPECIFICATION STRUCTURE

7.4.1 Scope

The scope section starts off with the full name assigned to the system covered. Immediately following is a brief description of what the system must accomplish, or a description of the larger system of which it is a part. The operational relationships to contiguous systems of the larger system should be described. Then the relationship to criticalness should be noted. Is the system critical to normal operation or to safety? Finally, the cognizant regulatory authorities should be named. What governmental agencies have jurisdiction? What insurance underwriters? Who else? This section delineates, in a general way, the complete range of work to be accomplished and the sources of the constraints on system design.

7.4.2 Applicable Documents

As usual, this section starts out with a standard statement concerning applicability of the documents listed therein. Then come documents originating in the specification writer's organization. These would be drawings controlling systems design, such as building drawings setting space limitations, or specifications or drawings concerning the defined interfaces with contiguous systems. Next come the codes and standards, as required by law or the customer, governing system design.

7.4.3 Design Requirements

The first major subsection under design requirements is design criteria. For a system specification these are the applicable codes and standards, and the details of their specific application to the system. The specification writer's organization may also want to apply some criteria.

The second major subsection concerns steady-state performance requirements for the system. Topics that would or could be included here are the system inputs and outputs, definitions of operating points, and the operating parameters with their associated ranges. Of these, the selected parameters used for controlling operation should be identified, and their set points and limits should be specified.

The third major subsection concerns transient performance requirements. Topics here would include such things as required or limiting rates of change between operating points and the primary control parameters—that is, those parameters that are directly controlled to produce changes in operating conditions. This subsection should also indicate which controls are to be automatic and which are to be manual. Included would be any requirements for manual overrides on automatic controls.

The fourth major subsection concerns steady-state structural requirements. Here, the topics covered are system loadings, expected load combinations (including anticipated "worst case" conditions), stress and deflection limits, environmental conditions (including both the normal operating environment and any postulated abnormal environment such as that caused by failures or accidents), and the required service life.

The fifth major subsection contains the transient structural requirements. The initial topic should be transients expected in the system loadings. Following that should be a definition of the coincidence of steady-state and transient loads. Expected transients in the environmental conditions should be covered if they are significant. Finally, limitations on or evaluation of cyclic effects and fatigue life should be specified.

7.4.4 Safety Requirements

The safety criteria go into the first major subsection under safety requirements. As before, the topics are the applicable codes and standards defining and imposing safety requirements and the details of their specific application to the system. Most frequently, these would be federal standards and insurance underwriters' codes.

The second major subsection defines all known possible unsafe conditions against which safety precautions must be taken. The topics could be arranged into events or conditions caused by abnormal steady-state operation and abnormal transient operation. Consideration should be given to all equipment failures in the process system and control system that could cause abnormal operation.

The third major subsection specifies required safety features. A number of these may be determined by the codes and standards; others may be required by a customer or the specification writer's organization. Topics would include the parameters used and perhaps controlled by the safety system, the details for its operation, and any specifically required components. Wherever critical safety functions exist, standard, tried, and proven components are usually used to ensure reliability. New safety component designs must first build up a satisfactory operating history in noncritical applications.

7.4.5 Operational Requirements

Under operational requirements, the first major subsection defines operating modes. Topics should include startup, normal operation, normal shutdown from a set of normal operating conditions, and abnormal shutdown from an abnormal or accident condition. This subsection depends very largely upon the type of system covered.

The second major subsection concerns operating equipment. Topics would include the desired automatic and manual controls. Display of indicated parameters should be specified and requirements for recording or simple indication should be defined.

The third major subsection concerns operating locations. Topics include items such as requirements for a main control station, redundant controls, and control arrangement. Control arrangement includes separation of main and redundant control functions, as well as spatial arrangement of control panels for the different control functions.

The fourth and final major subsection would cover requirements for system-operating personnel. This will directly affect the design of operating equipment. If the size of the operating crew is limited, the maximum size should be stated. Functions of different members of the operating crew should be stated, if this can be determined. This will affect the layout of controls and control equipment.

7.4.6 Maintainability Requirements

The first major subsection under maintainability requirements is for inspection-related requirements. The first topic would be requirements for accessibility to components that will be inspected regularly. Other topics would be the methods and equipment to be used for inspection. The system designer must know this in order to make the necessary provisions.

The second major subsection is concerned with maintenance requirements. Topics covered here will directly affect the time and cost necessary to maintain the system. The first is limitations on removal of equipment for maintenance. Requirements for in-place maintenance should be stated. Related to this are space requirements for maintenance equipment and personnel. Then there are requirements for replaceable components and component parts in the system and its equipment. When should complete items of equipment be replaceable, and when should their parts be removable and replaceable without removing the main structure? Directly related are requirements for interchangeability of parts between similar items of equipment. Finally, limitations on shutdown of the system for maintenance should be covered. Partial operation during maintenance may be a requirement.

The third major subsection is concerned with removal and replacement of components of the system if necessary. The topics here are mainly limitations on the tooling to be used and on handling or handling fixtures. This is directly related to the space requirements of the previous subsection.

7.4.7 Documentation Requirements

The documentation requirements of a systems specification should be very comprehensive. Quite naturally, the first major subsection is for drawings. This subsection lists all drawing types required for complete definition of the system. The purpose of this is to get the system designer to produce a drawing structure that matches that of the specification writer's organization.

Logically following are requirements for the associated lists. Of primary importance is a master parts list. Separate sublists, such as a pump list, a valve list, a motor list, or an instrument list, may be required. A separate list may be required for all components that will eventually require their own design or purchase specifications. Still another list may be required for all standard components that will require purchase specifications. Depending upon how the specification writer's organization wants to handle things, these lists may be required to reference other documents for design information, or they may contain all pertinent design information.

The third major subsection concerns the performance design report, which defines all component functions, operating parameters, and their levels, which, when combined, produce the desired system outputs for the given inputs. This report should not only show that the system will work; it should also show that all component requirements are within reason.

The fourth major subsection concerns the structural design report. This report contains all the stress and deflection analyses accomplished in support of the design. Any studies of the effects of cyclic operation and fatigue would also be reported here. If the codes and standards governing the design have required formats and contents for this report, they would have to be taken into account.

The fifth major subsection contains requirements for preliminary operating instructions or procedures. These will reflect the requirements of section 5, and will contain the details dependent upon the finished design. The extent of requirements for these details should be specified.

The sixth major subsection, in a like manner, contains requirements for preliminary maintanance instructions or procedures. These will reflect the requirements of section 6. Once again, the extent of requirements for details should be specified.

7.5 SYSTEM TEST SPECIFICATIONS

Only a minor departure from the component test specification structure is required here. The first-level structure of a system test specification consists of the following seven sections:

1. Scope
2. Applicable Documents
3. Acceptance Criteria
4. Test Conditions
5. Test Instrumentation
6. Test Procedures
7. Test Documentation

A comparison of this with the structure for a component test specification reveals the differences resulting from the changed nature of the task. The objectives, or test criteria, have been replaced by the acceptance criteria and the test conditions under which these criteria must be met. The differences will become clearer in the following section of this chapter.

There is no section for the test set-up. The system itself is the test set-up.

7.6 INITIAL ASSUMPTIONS

A large, field-assembled system must undergo considerable preparation before it can be tested as required by a system test specification. These preoperational checks and tests are presumably complete before testing is begun. They may be considered as the initial assumptions for the system test specification, and are as follows:

1. All postinstallation inspections, checks, and tests of all components have been successfully completed.
2. All operational connections (power, control, instrumentation, and so on) have been made.
3. All preoperational structural tests are complete.
4. All circuit verification (continuity) checks are complete.
5. All initial (standby) conditions have been set.
6. All initial startup safety precautions have been taken.

When these assumptions have been satisfied, the system is ready for testing to the requirements of a system test specification in accordance with its derivative instructions and procedures.

7.7 SYSTEM TEST SPECIFICATION STRUCTURE

7.7.1 Scope

As usual, the scope section sets the specification in perspective. It covers such items as the system name, what it is used for or what larger system it is part of, and its operational relationship to other systems. This discussion should be brief. The scope should then identify all types of tests to be performed on the system. Examples would be steady-state operational tests, transient tests, and safety function tests. Finally, any supervisory authorities who must oversee the testing should be identified.

7.7.2 Applicable Documents

This section, as usual, begins with a standard statement, usually concerning conflicts between the listed reference documents. The first documents listed are the system design documents. These include drawings, such as system schematics and assembly drawings, and specifications for the system and its components.

The next listing is of applicable operational documents. These include special operation manuals, written for the system or its subsystems, and new components designed especially for the system. Also included are standard manuals for the system's standard components.

The final listing is for applicable codes and standards governing testing of the system. These should be listed in order of precedence, if such an order exists.

7.7.3 Acceptance Criteria

This section defines the basis upon which acceptable operation of the system is determined. It includes a specification of performance parameters, measured and calculated, which will be used to determine acceptable operation. This is particularly important when the system involved consumes fuel or power and must meet efficiency guarantees, as well as a minimal output.

7.7.4 Test Conditions

This section may contain considerably more conditions than those under which the acceptance criteria must be met. Test conditions related to acceptance criteria may be given separately or may be included with the others and identified as they occur. Test conditions specified may include steady-state operating limits (extreme operating points), steady-state design points, and any additional steady-state points of interest for system operation.

System transient operation may also be a concern. The specified test conditions may include transient operating limits (as set by the system controller), transient design points (if process rates are important), and any additional transient operation points of interest for stable system operation.

Test conditions may be included for system switching and transfer points. These are the points at which the system switches (under automatic control) or is switched (under manual control) from one mode of operation to another, and at which operations are transferred from one component to another or even to another system.

Finally, test conditions may be specified to bring various safety features into operation and verify their ability to perform their functions when installed in the system.

7.7.5 Test Instrumentation

For many system tests the standard system instrumentation is not sufficient. This is particularly true when highly accurate measurements are required for the acceptance criteria. The first items under test instrumentation, then, are for the additional test measurements required. After that come requirements concerning methods of recording the data and accuracy requirements. For the standard instrumentation, these requirements are limited by what is available. Calibration requirements should indicate whether traceability to the National

Bureau of Standards is required. Finally, requirements for any special monitoring of safety functions should be covered.

7.7.6 Test Procedures

These procedures are closely related to manuals provided for operation of the system and its components. There should be procedures for standard (normal) system operation, such as startup, minimum output, maximum output, and shutdown. Then there should be procedures for any special initial system operation, such as verifying the correct settings of adjustable set points and limits in the system's controls. The required order of these operational tests should be indicated clearly.

7.7.7 Test Documentation

The final section is concerned with requirements for documentation resulting from the system testing. A system acceptance report should be the first item covered, detailing how the acceptability of the system has been demonstrated. It may also indicate where and how the system has failed to meet requirements. Perhaps a separate general report is required, much broader in scope than the acceptance report, covering all aspects of the testing. Something may be said concerning the final form and disposition of all the test data. They may go into an appendix of the report, or they may be handled in some other manner. The same, or something similar, may be said concerning calibration records or certificates. Another important item, which may or may not go into the report, is the record of all final settings and adjustments made to the controller and the final control elements. This will be required by future operations and maintenance personnel. One further item may be included, which is often taken for granted: the recommendation for modifications to the initial operating procedures, based upon the testing experience. These may not be made available unless they are explicitly required.

Part III

Specification Microstructure

Chapter 8

Specification Users and Use

8.1 THE SPECIFICATION ANALYST

Specifications are written for the equipment supplier. They tell the supplier what the equipment must be able to do. On government projects they may also be written to satisfy the bureaucracy overseeing the work. Even so, the needs of the equipment supplier must take precedence. Within a supplier's organization are one or more specifications analysts, even though they may not have that particular title. They are the key personnel involved in the use of specifications; it is to them that specifications are directly addressed.

8.2 THE ANALYST'S WORK

The basic function of the specifications analyst is to fit all of the specification's requirements into the system his or her company uses for producing their equipment. The analyst must determine which requirements will be handled by which of the various groups in the

71

organizational structure. He or she must direct the attention of each group to the requirements for which it will be responsible. This is often done with written instructions directed to each group, identifying the requirements that apply to its work. This may also be accomplished with annotated copies of the specification, if it is suitably written. The analyst may also have to provide some direction to the groups for integration of their work and for handling intergroup conflicts over interpretations of the requirements.

8.3 FACILITATING THE ANALYST'S WORK

The specification writer should keep the task of the specifications analyst constantly in mind. Requirements should be easy to locate and identify. This is accomplished by following a logical structure and leaving out excessive descriptive material. Requirements should be easy to understand, expressed in simple declarative sentences. The analyst will want to sort out requirements for the different working groups in his or her organization by marking up the specification. If this can be done with margin markings only, the analyst's work is much easier. Setting the requirements down in a list format when possible will facilitate this. Providing the analyst with clear, concise requirements will reduce considerably the amount of discussion and explanation required prior to contract negotiation.

Chapter 9

Specification Language

9.1 THE BASIC REQUIREMENT

The following quotation is taken from Military Standard 490 and the Department of Defense Standardization Manual:

> The paramount consideration in a specification is its technical essence, and this should be presented in language free of vague and ambiguous terms. Using the simplest words and phrases will best convey the intended meaning. Inclusion of essential information shall be complete, whether by direct statements or references to other documents. Consistency in terminology and organization of material will contribute to the specification's clarity and usefulness. Sentences shall be as short and concise as possible. Punctuation should aid in reading and prevent misreading. Well-planned word order requires a minimum of punctuation. When extensive punctuation is necessary for clarity, the sentence(s) shall be rewritten. Sentences with compound clauses shall be converted into short and concise separate sentences.

This quotation is directly related to the work of the specifications analyst. The analyst must not be confused or misled by the way the requirements are stated. He or she must be able to recognize them immediately and see their relationships clearly. This is difficult to do if the specification contains multiple requirement, involved (although grammatically correct), paragraph-long sentences.

9.2 IDENTIFYING SPECIFIED REQUIREMENTS

The verb form "shall" is used for requirements placed upon the supplier—that is, for things the supplier shall do, documents he or she shall supply, features he or she shall build into the equipment, performance levels the equipment shall meet, and so on. Wherever the word "shall" appears, it indicates that a requirement is being stated.

Example 1

The supplier shall prepare a design report and shall submit it to the purchaser for review and approval at least 60 days prior to start of fabrication.

Example 2

Gearing shall be grease-lubricated. The gearbox shall have grease fittings for replenishing the lubricant, and shall have an access cover for maintenance inspection.

Example 3

Terminal blocks shall be 600 volt, front-connected, barrier-type, with sliding links and marker strips identifying all internal and external wiring.

Example 4

Removal of any logic card shall be indicated on the system diagnostic display.

The verb form "will" is used for requirements the purchaser is placing upon himself or herself—that is, for information the purchaser will supply, documents he or she will review, approvals he or she will issue, and so on, all at the proper time.

Example 5

The purchaser will review material certifications and issue the release for fabrication within two weeks after the final submittal.

Example 6
The purchaser or his designated representative will witness the final acceptance test and approve the shipping release.

Example 7
The purchaser will supply the final nozzle loads within 60 days after receipt of the outline drawing showing nozzle locations and sizes.

Example 8
The purchaser will review all safety logic schematics prior to circuit board design release.

Note that for requirements the future tense is used. The present tense is used for giving directions.

Example 9
Specific design and performance parameters for each item are listed in the data sheets.

Example 10
Qualification requirements unique to a valve type are given in the type specifications.

Present tense is also used for indicating options.

Example 11
Either carbon steels or low-alloy steels are acceptable for structural components.

Example 12
Available power sources are either 480 VAC 3-phase or 120 VAC 1-phase.

However, options may also be stated in standard requirement form.

Example 13
Structural components shall be either carbon steel or low-alloy steel.

Example 14
The component shall be operable from either a 480 VAC 3-phase or a 120 VAC 1-phase power source.

Either form is acceptable. The writer should follow his or her organization's standard practice, or be consistent throughout the specification.

9.3 THE DEFINITE APPROACH

Avoid using the word "should" wherever possible. It is to indefinite and normally implies a preference not strong enough to warrant the use of the word "shall." Also, it is often used to introduce or imply an option where the alternative(s) are not stated. It is possible to introduce preferred alternatives and allowable alternatives with the definite "shall," and this is the best practice.

> *Example 1 (undesirable)*
> Valve stems should be austenitic stainless steel.

What are the possible alternatives? Martensitic steel? Precipitation-hardened steel?

> *Example 2 (preferred over example 1)*
> Valve stems should (shall) be either austenitic or martensitic steel.

The above, however, does not indicate an order of preference.

> *Example 3 (preferred over examples 1 and 2)*
> Valve stems shall be either austenitic steel (preferred) or martensitic steel (acceptable).

Examples 2 and 3 have eliminated precipitation-hardened steels from consideration, and example 3 has made the order of preference clear. Clearly specified alternatives are an advantage to both the supplier and the purchaser.

> *Example 4 (undesirable)*
> Logic noise threshold should be greater than 10 volts, and buffer isolators should isolate digital instrument channel voltage from the logic system voltage by about 2,500 volts or more.

Why should it be greater than 10 volts? Why not 8 volts, or even 12 volts? What signal-to-noise ratio should be considered when determining the actual value? Why 2,500-volt isolation? Wouldn't 2,000 volts be adequate and cheaper?

> *Example 5 (preferred)*
> Logic noise threshold shall be greater than 10 volts, and isolation of digital instrument channels from the logic system by buffer isolators shall be 2,500 volts or more.

The supplier may then assume the signal-to-noise ratio and the cost of isolation limits have been considered.

Example 6 (undesirable)
Low-level signal wiring shall be separated from high-level signal and AC power wiring by adequate spacing or other standard means.

What spacing is considered adequate? What other standard means does the writer have in mind?

Example 7 (preferred)
Low-level signal wiring shall be separated from high-level signal and AC power wiring by a minimum spacing of three feet, or by the use of ferromagnetic conduit.

Examples 5 and 7 leave no questions concerning the requirements or alternatives. However, example 7 gives the supplier the option of deciding which wiring to place in conduits. This is acceptable as long as the writer is aware of it.

Specification writers sometimes explain or give reasons for their choices of alternatives. A standard rule in specification writing is to state requirements only and to give no reasons for them.

Example 8 (undesirable)
In order to minimize stress corrosion cracking, valve stems shall be austenitic or nonsulfurized martensitic stainless steels. Nonsulfurized martensitic steels are required to minimize the possibility of torsional failure.

Example 9 (preferred)
Valve stems shall be austenitic or nonsulfurized martensitic stainless steels.

Example 10 (undesirable)
Due to the expected equipment vibration, terminal block lugs shall be screwed, ring-type, with lockwashers, which will prevent them from falling off if the screws loosen.

Example 11 (preferred)
Terminal block lugs shall be screwed, ring-type, with lockwashers.

Example 12 (undesirable)
To avoid ground loops, shields connected to the ground bus shall not be connected at both ends. Connections shall be at the panel end only. DC

power for 4–20 ma control loops must not be connected to the panel ground.

Example 13 (preferred)
Shields shall be connected to the ground bus at the panel end only. Control loop 4–20 ma DC power shall be separately grounded.

The supplier may disagree with your reasons and use them as an opening for proposing further alternatives where none are wanted. The purchaser and the supplier usually will get together in at least one fact-finding session anyway, and questions concerning reasons for requirements can be discussed then. Still another consideration is the fact that there may be more than one reason for a requirement. Thus, the requirement may stand while the reason for it changes. Giving reasons for requirements, then, may lead to revisions that could have been avoided.

Too much descriptive material also should be avoided. It tends to obscure the requirements.

Example 14 (undesirable)
In addition to its normal operation transients, this component may undergo a number of abnormal transients due to system failures or unplanned events, such as loss of power, unexpected severe environmental changes, or operator error. The component shall be operable after all of these transients, both normal and abnormal, which are tabulated in the data sheets.

Example 15 (preferred)
The component shall be operable after all normal and abnormal transients tabulated in the data sheets.

A good specification is limited to stating its requirements in definite form, with a minimum of other material.

Chapter 10

Stating Requirements

10.1 THE SIMPLE DECLARATIVE SENTENCE

The usual form for stating requirements is the simple declarative sentence. State the requirements as facts, and nothing else, whenever possible. Do not carry this to the point where the writing becomes stilted and unnatural. All requirements can be stated this way, but not all requirements should be.

Example 1 (preferred)
Actuators shall be single-acting, spring-return, either piston or diaphragm type, with integral air sets and pilot valves.

Example 2 (undesirable)
Actuators shall be single-acting, spring-return. They shall be piston or diaphragm type. They shall have integral air sets and pilot valves.

Example 3 (preferred)
Load drivers shall change state on a 100-microsecond test pulse, and return to normal in less than 9 milliseconds.

Example 4 (undesirable)
Load drivers shall change state on a 100-microsecond test pulse. Then they shall return to their normal state in less than 9 milliseconds.

10.2 MODIFYING PHRASES

Many requirements are not totally applicable to the equipment covered. They must be modified for some areas. The simple declarative sentence is not used if a modifying phrase is needed. Restricting the range of application of a requirement is a common example.

Example 1
Coded structural components shall be low-alloy steel when the operating temperature is in the 700°-to-1,100°F range.

Example 2
When the operating temperature is in the 700°-to-1,100°F range, coded structural components shall be low-allow steel.

Example 3
Motors shall be capable of withstanding overspeeds 25 percent above synchronous speed when rated at 1,800 rpm and below.

Example 4
When rated at 1,800 rpm and below, motors shall be capable of withstanding overspeeds 25 percent above synchronous speed.

Examples 1 and 3 are preferred, and examples 2 and 4 are acceptable. The important point is for the writer to be consistent in sentence structure throughout the specification.

10.3 ADDING EXCEPTIONS

In addition to basic requirements and modifications, there may be exceptions. Exceptions may be placed in the sentences containing the requirements as clauses (especially when there are no modifications), or placed in follow-on sentences.

Example 1
Coded structural components shall be low-alloy steel, except for the Class D items.

Example 2

Coded structural components shall be low-alloy steel when the operating temperature is in the 700°-to-1,100°F range. Class D items are excepted.

Example 3

Except for the Class D items, coded structural components shall be low-alloy steel.

Example 4

Except for the Class D items, coded structural components shall be low-alloy steel when the operating temperature is in the 700°-to-1,100°F range.

Example 5

Coded structural components having operating temperatures in the 700°-to-1,100°F range, except for the Class D items, shall be low-alloy steel.

Example 6

Motors shall be capable of withstanding overspeeds 25 percent above synchronous speed, except for fan-drive motors.

Example 7

Motors shall be capable of withstanding overspeeds 25 percent above synchronous speed when rated at 1,800 rpm and below. This requirement does not apply to fan-drive motors.

Example 8

Except for fan-drive motors, motors shall be capable of withstanding overspeeds 25 percent above synchronous speed.

Example 9

Except for fan-drive motors, motors shall be capable of withstanding overspeeds 25 percent above synchronous speed when rated at 1,800 rpm and below.

Example 10

Motors, except for fan-drives, when rated at 1,800 rpm and below, shall be capable of withstanding overspeeds 25 percent above synchronous speed.

Examples 1, 2, 6, and 7 are preferred; examples 3, 4, 8, and 9 are acceptable; and examples 5 and 10 are undesirable. Once again, the writer should be consistent with sentence structure throughout the specification. When a specification contains a variety of sentence structures and styles, it frequently indicates that the writer copied requirements from several other specifications without doing any rewriting.

10.4 MULTIPLE APPLICATIONS

There may be times when a requirement applies to a number of components or parts. In this event a listing of the items is the preferred method for giving the requirement. State the requirement and then list its applications. If, for a particular item in the list, there are exceptions, place the exceptions in parentheses after the item.

Example 1
The following parts shall be made from austenitic or nonsulfurized martensitic stainless steel.
   a. globe and gate valve stems
   b. check valve axles
   c. gland flanges
   d. gland bolting
   e. lantern rings (except for bellows-seal-type valves)

Example 2
The following components shall be operable from a 120-Vac power source:
   a. fan-drive motors
   b. louver actuators
   c. hydraulic pumps
   d. machine controls
   e. pilot valve solenoids

Chapter 11

Organizing Requirements

11.1 LIMITS ON LEVELS

The decimal numbering system for sections, subsections, and paragraphs permits an indefinite division into subheadings. Limiting the number to four, however, is good practice. Department of Defense Manual 4120.3-M calls for a limit of three, but four is the more generally accepted limit. Holding to this limit requires more thoughtful arrangement of the material. The result of the effort is always a better specification structure.

In Part II of this book the first three levels of organization of an engineering specification were discussed. The first level is usually hard and fast in any organization. The second level is based on a logical structure and should be considered more flexible. The third level should be looked upon as a suggested logical organization only.

Now we will consider organizations of the specific requirements at the bottom or final levels, requirements that will be written in accordance with chapters 9 and 10 of this book. The main concern will be to organize them into final paragraphs and lists.

11.2 SEPARATING REQUIREMENTS

Do not include several different requirements in one continuous paragraph. Use as many paragraphs under a next higher level heading as necessary to provide one for each definable requirement, or place these requirements in a list. Listing of requirements will be considered in detail later. This section will deal with paragraphs.

Example 1 (undesirable)
3.3.2 *Piping.* All piping shall be stainless steel, with a minimum size of ¼-inch OD tubing, having a wall thickness of .035 inches. Piping connections shall be female NPT with elastomeric O-ring seals, and flareless, compression-type, stainless steel connectors. The supplier's standard maintenance shutoff valves shall be provided as necessary.

Example 2 (preferred)
3.3.2 *Piping.* All piping, fittings, and connectors shall be stainless steel and shall meet the following requirements:
3.3.2.1 Connections shall be female NPT with elastomeric, O-ring seals.
3.3.2.2 Connectors shall be flareless, compression-type.
3.3.2.3 Minimum size shall be ¼ OD tubing, with a .035-inch wall thickness.
3.3.2.4 The supplier's standard maintenance shutoff valves shall be provided as necessary.

Example 3 (undesirable)
3.3.5 *Enclosures and Conduit.* Enclosures and conduit for wiring, terminal blocks, switches, etc., shall be NEMA Type 4 with access covers. Conduit between fixed enclosures shall be rigid steel, hot-dipped galvanized inside and out, and that from movable enclosures shall be flexible, with insulated connectors. Continuity of ground shall be maintained across flexible conduit connections, with the wiring installation in accordance with the National Electrical Code NFPA 70.

Example 4 (preferred)
3.3.5 *Enclosures and Conduit.* Wiring, terminal blocks, switches, etc., shall be installed in enclosures and conduit meeting the following requirements:
3.3.5.1 Enclosures shall be NEMA Type 4 with access covers.
3.3.5.2 Conduit between fixed enclosures shall be rigid steel, hot-dipped galvanized inside and out.
3.3.5.3 Conduit from movable enclosures shall be flexible, with insulated connectors.
3.3.5.4 Continuity of ground shall be maintained across flexible conduit connections.
3.3.5.5 Wiring installation shall be in accordance with the National Electrical Code NFPA 70.

The examples here are short and simple. In many instances such requirements would be longer and more complex.

11.3 MULTIPLE SUBSECTIONS AND PARAGRAPHS

There will be times when a writer gets down to the last, or next to the last, decimal-numbered level, and still has material in quantity and variety that cannot be organized without exceeding the level limit. The problem can be solved by the use of multiple subsections, paragraphs (with lists), or both. This involves placing material on the same subject in several consecutive subsections or paragraphs at the same level.

Preferably, the paragraphs should be independent of each other. If they cannot be made independent, the first one defines their relationship and states its application to the following subsections or paragraphs. When all material within this purview has been covered, the specification format goes up at least one level. This indicates that the first controlling subsection or paragraph no longer applies. Use of this dependent form of the alternative described here requires that the series of equal-level subsections or paragraphs be placed at the end of its own next higher level.

Complete examples of the above would be excessively long. Therefore, skeleton examples are provided, which should be sufficient to illustrate the application of this solution.

> *Example 1 (independent)*
> 3.3.3 Design Selection Criteria for Drives
> 3.3.4 Pneumatic Mechanisms
> 3.3.5 Pneumatic-Hydraulic Mechanisms
> 3.3.6 Electrohydraulic Mechanisms
> 3.3.7 Electric Motor Mechanisms

Here, the first subsection gives criteria for making a selection from among four types of drive mechanisms. It does not place any requirements on the types themselves. Once a type has been selected, its requirements are completely covered by one of the other subsections. In this manner all five subsections are kept at the same level, thus providing an additional level for use under the different types of drives.

> *Example 2 (dependent)*
> 3.3.3 General Requirements for Drives
> 3.3.4 Pneumatic-Hydraulic Mechanisms
> 3.3.5 Electrohydraulic Mechanisms
> 3.3.6 Electric Motor Mechanisms
> 3.4 Materials Selection

Here, the first subsection gives requirements applicable to all types of drive mechanisms, without exception, and is the controlling subsection.

The next three subsections give particular requirements for each of three types. If one of these types is selected, the applicable requirements are those of its subsection plus those of the first subsection. If some other type is selected, the applicable requirements are those of the first subsection only.

The level change at the end indicates termination of the range of application of the first subsection. In this instance the title itself may also make it obvious.

Sometimes a dependent group of subsections or paragraphs is allowed to occupy a position other than the end of its own next higher level. Then the first (controlling) subsection or paragraph must identify its range of application by the numbers. The main reason for not using this approach is the possibility of error, after extensive revision, if the numbers are inadvertently left unrevised.

11.4 THE USE OF LISTS

Alphabetically itemized lists and numerically itemized sublists may be used anywhere after the second, third, or fourth levels have been reached. This results in an effective maximum of six levels of arranging the material.

Ideally, the items in a list should require one line each. They should be singular entries of very restricted scope.

Example 1
a. Shafts shall be austenitic stainless steel.

Example 2
b. Terminal block connections shall have ring-type lugs.

However, if identifying exceptions or contingencies is necessary, they may run for two or three lines.

Example 3
a. Valve bodies shall be carbon steel except when the design temperature is above 700°F.

Example 4
b. Wiring shall be stranded copper No. 14 AWG with crimp-type terminals unless otherwise specified in the data sheet or the applicable standard.

Excessively long items in lists sometimes represent an attempt to evade the limit on decimal-numbered subsections. At other times they represent a packing in of additional material from reviewers' comments. Restructuring the list is preferable.

Lists may be used for separating requirements without resorting to a large number of short, decimal-numbered items. This also avoids long, multirequirement paragraphs.

Example 5 (undesirable)

3.5.1 *Design Report.* The supplier shall submit a design report for the purchaser's review and approval. The report shall contain the results of calculations performed in support of the applicable code requirements. It shall also contain an analysis for the additional loading conditions specified above, and for fatigue effects due to the transients specified in the data sheet. Finally, it shall contain the calculations for the performance characteristics specified in the data sheet. The analyses shall contain sufficient detail to permit checking of the results.

Example 6 (preferred)

3.5.1 *Design Report.* The supplier shall submit a design report for the purchaser's review and approval. The report shall consist of the following:

   a. Results of calculations performed in support of the applicable code requirements.
   b. An analysis for the additional loading conditions specified above.
   c. An analysis for fatigue effects due to the transients specified in the data sheet.
   d. Calculations for the performance characteristics specified in the data sheet.

The analysis shall contain sufficient detail to permit checking of the results.

Example 7 (undesirable)

5.1.4 *Vibration Test.* Vibration testing shall be accomplished with 1.5 g level uni-axial sine sweep runs, through the range of 4 to 64 Hz. The sweep rate shall be 1 octave per minute, and the g level measured at the base accelerometer. Three runs shall be made: one up-and-down range sweep for each of three orthogonal axes, two of which shall be in planes of symmetry of the equipment.

Example 8 (preferred)

5.1.4 *Vibration Test.* Vibration testing shall be accomplished in accordance with the following requirements:

   a. Test type: uni-axial sine sweep
   b. Acceleration level: 1.5 g at the base accelerometer
   c. Frequency range: 4 to 64 Hz
   d. Sweep rate: 1 octave per minute

  e. Test runs: one up-and-down range sweep for each axis

  f. Test axes: three orthogonal axes, two of which shall be in planes of symmetry of the equipment

In most instances, lists of requirements will take up more space than paragraphs. However, they obviously make it much easier for the specifications analyst, or anyone else, to sort out the individual requirements into whatever groups he needs for his purpose.

  Also note that lists may occur at the end of a paragraph or within it, as shown in example 6. In fact, more than one list may be placed in a paragraph.

Example 9 (undesirable)

3.3.4 *Analog Input Channels.* Channels for analog input signals shall operate at 4-20 ma from a 24 VDC supply. The channels shall be tripped by a 0 to 15 volt comparison with accuracy greater than 25 percent of set point, and have a fail-safe design. They shall have interference filters with an actuation response time delay no greater than 20 milliseconds. They shall also have meters on the front panel for testing and calibration.

Example 10 (preferred)

3.3.4 *Analog Input Channels.* Channels for analog input signals shall meet the following requirements:

  a. Signal range of 4-20 ma

  b. Power supply of 24 VDC

  c. Trip level of 0 to 15 volt comparison

  d. Accuracy greater than 25 percent of set point

Channel circuitry shall incorporate the following features:

  e. Interference filters with an actuation response time delay no greater than 20 milliseconds

  f. Front-panel meters for testing and calibration

  g. Fail-safe design

This approach offers more flexibility than using a series of decimal-numbered items. However, this should not be overdone as a way of avoiding careful organization into several paragraphs. A limit of three short lists per paragraph should be observed. This is another situation where the writer must examine his or her material and use judgment.

11.5 SUBLISTING

Sublisting is an extension of the method of reducing requirements to their simplest elements and giving these elements maximum visibility.

It may be used for separating groups of related requirements or groups of options, and for further reducing main list items.

Example 1 (undesirable)

a. Switches shall be cam-operated, double-pole-double-throw, adjustable, and capable of being locked in the position selected.

b. The contacts shall have a wiping action when closing, shall be silver-plated and rated for 10 amperes at 120 VAC resistive or 0.5 amperes at 125 VDC inductive.

Example 2 (preferred)

a. Switches shall have the following design features:
   1. cam operation
   2. double-pole-double-throw action
   3. adjustable position (with position lock)

b. Switch contacts shall have the following design features:
   1. wiping action when closing
   2. silver-plating
   3. rating, alternative 1: 10 amperes at 120 VAC resistive
   4. rating, alternative 2: 0.5 amperes at 125 VDC inductive

Example 3 (undesirable)

a. Welding shall be accomplished using the following processes: shielded metal-arc, gas tungsten-arc, submerged arc, gas metal-arc, and flux-cored arc, with hard-surfacing operations also using oxyacetylene and plasma-arc methods.

Example 4 (preferred)

a. Welding processes used shall be:
   1. shielded metal-arc
   2. gas tungsten-arc
   3. submerged arc
   4. gas metal-arc
   5. flux-cored arc

b. Hard-surfacing operations shall also use:
   1. oxyacetylene
   2. plasma-arc

Sublist items should always be held to one line and do not have to be complete sentences. They should state a single, specific requirement or fact, with no exceptions or qualifications beyond brief parenthetical expressions such as "if applicable," "as applicable," "as required," "if available," and so on.

Sublisting is useful for adding modifications to a requirement, such as for particular ranges of parameters. Sublists are also useful for adding a list of exceptions to a requirement. Adding to and deleting from such sublists are very simple tasks.

Example 5
c. Envelope dimensions
   1. hardware envelope
   2. removal space requirements
   3. maintenance fixture locations (if applicable)
   4. total maintenance space envelope (as required)

Example 6
d. The maximum thickness of each layer for submerged-arc welds shall not exceed the following:
   1. ½ inch for steels with a thickness of 1¼ inches or more
   2. ⅜ inch for steels with a thickness less than 1¼ inches
   3. ¼ inch for austenitic stainless steels of all thicknesses

Example 7
e. The following materials are disallowed:
   1. halogen-containing materials
   2. copper-containing material
   3. nitrided surfaces in contact with the process fluid
   4. sulfur and low-melting-point metals or their compounds

11.6 MULTIPLE REQUIREMENT ENTRIES

Finally, there are acceptable exceptions to the rules given above. A group of parameters that frequently appear together may be stated as if they were a single requirement, or as a single item in a list. This usually occurs for standard input or output parameters, such as power requirements, thermal hydraulics, and so on.

Examples
a. Power: 460 VAC, 60 Hz, 3-phase
b. Air conditions: 150 ± 10 psig, −40°F DP, 1.5 micron cleanliness
c. Inlet conditions: 25 psig min. at 450°F
d. Power output: 2,000 hp at 1,800 rpm
e. Design point conditions: 150 K-lb/hr at 3,500 psig and 200°F

11.7 FINAL COMMENTS

This part of the book has covered a variety of examples from the three major areas of design, manufacturing, and testing. The observant reader will have noticed that there is a difference among the three. Generally speaking, a writer will use the least number of lists in the design section and the most in the manufacturing section. It depends upon the level of detail in each. The writer should not strive to make each of these major sections look alike. He or she may strive to make all subsections in each look alike.

There are basic differences beyond the scope of specification writing. However, a specification writer, taking into account these differences, may still do a first-class job in writing a specification covering all three areas. This requires both a knowledge of specification-writing practices and good judgment. This book can supply only the first. Good judgment must be supplied by the writer.

Chapter 12

Titles and
Tables of Contents

12.1 TITLES

The first two levels of engineering specifications have hanging titles; there is nothing on the line after the title. The third and fourth levels have run-on titles; the first sentence of text begins on the same line after the title. Some organizations permit run-on second-level titles if the text following the title is a single introductory sentence. This practice can be useful for imposing a single code or standard on the entire section. The document may be imposed as a whole or "as indicated in the following paragraphs."

Titles are not required for the lowest-level paragraphs in a section regardless of their level. Lists may be used in untitled paragraphs or in place of untitled paragraphs, in and after the second level. Usually, the paragraphs of the scope section are untitled second-level paragraphs. The standard statements concerning applicability in the applicable documents section are usually left untitled.

12.2 TABLES OF CONTENTS

Tables of contents should list only the first two levels of a specification. In a well-organized specification this is all that is necessary to tell the user where to look for a particular requirement. If a user needs all titles listed through three or four levels to find what he or she wants, the specification is poorly organized.

Usually, the scope has only its first level (itself) listed, since its second-level paragraphs are untitled. However, in a long and complex scope, the second levels may be titled and listed in the table of contents. The applicable documents section usually is also listed by its first level only, because of the anomaly of its untitled standard statements. If the specification writer has a long list of codes and standards and wants to list their originating organizations in the table of contents, he or she must provide a title for the standard statements.

The table of contents is basically a guide to the contents of the sections containing requirements (from section 3 on). Any of the requirements sections may be listed by first level only if reference to a supporting specification or specifications are all it contains.

Part IV

Preparing Specifications

Chapter 13

The Process

The process of getting a specification ready for issue usually consists of the following stages.

The writer prepares the specification and has it typed. He or she then determines who the reviewers will be, according to the organization's rules, and has the appropriate number of copies made and sent out. The reviewers do their work and send the copies back to the writer, along with their comments. The writer evaluates the comments, resolves any conflicts, and has the specification revised to include the accepted comments or modifications. The writer then sends the specification to the editor. The editor performs the usual editorial tasks and also reviews the document for conformance to the organization's standards for specifications. The editor makes or has the writer make whatever modifications are necessary to obtain such conformance. The specification is then ready to be issued.

Some organizations have the editor do a preliminary edit of the document prior to the reviews. This reduces the incidence of reviewers doing editorial work. Other organizations do not have editors and expect the reviewers to do editorial work. This approach usually results in a poor quality of specifications, unless the organization has experienced writers and comprehensive, well-defined written standards.

There is usually a minimum of three reviewers, one to cover each of the major areas of design, manufacturing, and testing. Frequently, additional reviewers are assigned to concentrate on particular considerations, such as safety, reliability, code compliance, interfaces with other equipment, or cost control. Sometimes the writer is permitted to determine the number of reviews required beyond a specified minimum. Sometimes he or she must proceed according to a set schedule with no choice whatsoever. The larger the number of reviewers, the greater the chances of getting redundant or conflicting comments. The tendency of many reviewers to cover the entire document, not just their assigned area, contributes to this result.

Chapter 14

The Writer's Work

14.1 PRELIMINARY CONSIDERATIONS

Having looked briefly at the process of bringing a new specification into existence, we should now consider in more detail how a specification writer must approach the task of preparing a specification in a particular context—how the writer must consider the nature and ultimate use of the equipment involved. The ultimate use, of course, will be related to the branch of industry in which the writer is working. Each branch of industry reflects the major concerns of its customers, and those concerns in turn determine what that branch considers important in a specification.

The discussion up to this point has not considered what determines the relative complexity and variety of requirements in specifications for components or for a particular subsystem or system. Of the three or four major divisions of requirements, the design requirements are the ones that determine the complexity of a particular component, and thus most of the variety of tasks to be performed and the cost for delivery of acceptable items.

The single most important factor affecting the totality of design requirements (and their associated test requirements) is the distinction between passive and active components as discussed in the glossary. What makes activity important is the greater number and variety of design and associated testing requirements necessary for active items. This originates in the very nature of what is called *activity*. Active items are subject to failure from a greater variety of causes than are passive items. Simple aging and overloading are the primary causes of failure in passive components. For active components, wear, in a number of forms, becomes an important consideration. So also do a number of complex aging effects due to cyclic operation. These occur even when loading and operation are kept within normal limits. Environmental extremes usually have more adverse effects on active items. As a result, far more attention must be given to reliability and durability considerations for active components.

Along with this comes more involved maintenance concerns for active components. Many passive components, if properly used, require little or no maintenance at all. The main concern may be that they do not interfere with maintenance of active components. Finally, passive items usually can be made interchangeable more easily than active items. The importance of the distinction between passive and active items for the specification writer should now be clear.

Figure 14.1 shows the determinant considerations for design requirements (and associated testing requirements) for passive and active components, and how they fit into the subsections of a specification. A review of the glossary may be in order at this point.

For operational design requirements, passive components need identification and quantification of their defining parameters. In addition, active components need definition of interactive effects (from the model analysis) and of ranges of operation that affect a number of other components.

For structural design requirements and materials selection, passive components need consideration of reliability, durability, and safety. In addition, active components need consideration of the effects of excessive component interactions (from the model analysis) and of protection of interactive components.

For design feature requirements and materials selection, passive components need consideration of maintainability, interchangeability, and protection from adverse environments. In addition, active components need consideration of features for assuring reliability, durability, and survivability under adverse operation.

For the simplest passive components, design requirements may be placed almost entirely on a drawing. The design requirements section of the specification will then do little beyond referencing the

Figure 14.1 Design requirement determinants

SPECIFICATION SECTIONS	PASSIVE COMPONENTS AND SUBSYSTEMS	ACTIVE COMPONENTS AND SUBSYSTEMS
OPERATIONAL DESIGN REQUIREMENTS	1. The Defining Parameters and their Values and Limits	1. The Defining Operating Points which locate the curve or curves of operating parameters 2. The effects of off-design operation of interacting components on the above 3. The ranges which define the extent of operation
STRUCTURAL DESIGN REQUIREMENTS AND MATERIALS SELECTION	1. Expected Reliability 2. Expected Durability 3. Assurance of Safety	1. Expected Reliability 2. Expected Durability 3. Effects of excessive component interactions on the above 4. Protection of interacting components 5. Assurance of safety
DESIGN FEATURE REQUIREMENTS AND MATERIALS SELECTION	1. Ease of Maintenance 2. Range of Interchangeability 3. Protection from Adverse Environments	1. Features assuring reliability 2. Features assuring durability 3. Ease of maintenance 4. Range of interchangeability 5. Assurance of survivability under adverse operation 6. Protection from adverse environments

drawing and perhaps applying a code or standard. If the drawing covers a range of sizes, the sizing data may be tabulated on the drawing, eliminating the need for data sheets. This practice is most common at the final (lowest) level of specifications.

At this point further discussion of *reliability, durability, maintainability,* and *interchangeability,* known in military circles as "the ilities," is in order. Their relative importance depends upon the intended use of a particular equipment item. This has a strong influence upon how they are treated by the appropriate branch of industry and by the customer.

The military is concerned with all of the "ilities." This, of course, is because of the very demanding, uncertain, and crucial nature of battlefield use. Weaponry must function when needed; thus, reliability is important. It has to last under severe use throughout the battle; thus, durability is important. If some parts fail, they must be quickly and easily replaceable; thus, maintainability is important. If a number of similar weapons are damaged, a considerable amount of interchangeability will help get the largest number possible back into action. Concern for all of the "ilities" is greatest among contractors supplying equipment to the military.

Consider now the use of a weather or communications satellite sent into orbit, never to return. Reliability is of great importance; everything should function when and as the designer intended. Durability is next in importance. A device that wears out or degrades after six months, when it should have lasted for one year, is not acceptable. However, the loss is not as great as it would have been had it failed on initial startup. There is no intention or possibility of maintaining the orbiting satellite, so maintainability is of little importance. As for interchangeability, it may receive some attention if there are plans to use some of the same devices on future satellites. This is sometimes called *forward interchangeability.* However, interchangeability to facilitate maintainability is of no use.

Now consider flight equipment for commercial passenger aircraft. Reliability is important because it is so often required for safety. Critical equipment must not fail to perform during flight. Durability is important because it reduces maintenance and replacement costs, but it is not as important as in the satellite application. Maintainability is important for long and economical service life and for keeping up availability in the event of changing schedules. Interchangeability is of relatively low importance. With a network of repair shops, with stocks of spare parts and components, the replacement problem is minimal.

Finally, consider fixed installations on land, such as processing plants, or vehicles used in the construction industry. Durability is important because it minimizes long-term operating costs. Maintainability is also important because it contributes to lower costs and tighter scheduling of operations. Reliability is of lesser importance since safety is not often involved, and maintainability can make up for it. For many such equipment applications the approach is to concentrate on durability and maintainability, and take whatever reliability is available as technological "fall-out" from military and aerospace programs. Interchangeability is of little importance. In fact, an equipment manufacturer desirous of protecting a replacement parts market may want to avoid it.

From the above discussion it is evident that considerations of economics and safety have much to do with the relative importance attached to the "ilities" by various branches of industry and their customers. What all of this means to the specification writer is that a specification written for a general-purpose component will have some significantly different requirements depending upon the branch of industry ordering the item and its final use by the ultimate customer. A skillful specification writer will always take this into consideration.

14.2 SOURCES OF MATERIAL

Having considered how specification writers determine the magnitude of a particular task assigned to them, we should now turn to the sources of material and requirements the writer may use as he or she proceeds with the work of turning out complete specification drafts.

Figure 14.2 shows nine sources of material for a new specification. The three at the top represent flowdown of requirements from above, as was discussed in relation to Figures 3.1, 3.2, 3.3, and 3.4 in chapter 3. These, of course, are the necessary minimum.

In many instances the necessary minimum is not sufficient to completely define equipment covered by the new specification. The writer must then turn to other sources. On the left of the figure appear three other sources of information leading to additional requirements. The first of these, existing similar specifications, is most useful when they are available. This is the first additional source the writer should consult. Engineering handbooks can also be of assistance. Their treatment of various types of equipment will indicate the necessary design parameters and their relative importance and use. Industrial standards are also a useful source of information and guidance, particularly in the areas of manufacturing and testing requirements. Frequently, use of these documents will prevent the writer from specifying something quite logical but exceedingly difficult, or expensive, or both.

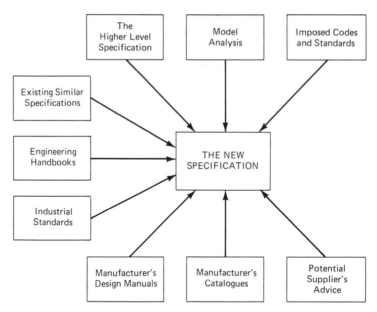

Figure 14.2 Sources of requirements

Finally, the writer can turn to organizations supplying the kind of equipment defined by the new specification, as shown along the bottom of Figure 14.2. Often, manufacturers publish design manuals or guides to assist potential users of their equipment in determining correct or optimal applications. These determinations are expected to appear in specifications or in data sheets for standard components and parts. Manufacturers' catalogues can be useful for simple components or parts where selection of sizes, ranges, and variable operating characteristics are important. Finally, the writer can seek advice from potential suppliers of the equipment being specified. Such advice will be forthcoming during the procurement process anyway, but sometimes it helps to get some of it in advance.

Judicious selection among and use of information from the six additional requirement sources discussed above will enable the writer to produce a reasonably good specification even though his or her initial familiarity with the specified equipment is limited.

14.3 REVIEWS AND EDITING

All specification writers want to turn out good documents, and well-taken comments from reviewers contribute to this goal. Good writers welcome good comments. They know that if one reviewer lets a flaw slip by, probably half a dozen bidders will call it to their attention. Working with reviewers who also want to see a good specification is easier than working with individual bidders who frequently are more interested in maintaining their competitive position. Worse still is the task of rewriting a specification during or after a bidders' conference.

No writer wants to undertake extensive rewriting and restructuring as a result of reviews and editing. To avoid this, the writer should know what to expect from the reviewers and the editor. He or she should use the writer's, reviewer's, and editor's checklist in this book while preparing a specification. In this way he or she can do the best job the first time, facilitate the reviewers' and editor's work, and hold down the number of comments received.

14.4 THE WRITER'S CHECKLIST

The following questions comprise the Writer's Checklist in appendix I. This discussion will define the scope of each question to enhance the new writer's understanding. More experienced writers will undoubtedly want to use the compact format in the appendix as a tool of their trade.

Have I used the correct format?   The required format may depend upon customer requirements or special internal organizational requirements as well as the specification type. If the writer is expected to know which one to use for a given set of circumstances, he or she is well advised to know just that. Having a draft specification sent back for reformatting is one of the least desirable experiences a writer can have.

Have I organized the division into subsections and lists in a logical manner?   The concern here is that the same logical structure has been followed throughout. If, for instance, the writer has included a separate treatment for each of several components, and the same or similar points are included for each, then the order in which these points occur should be the same for each.

Have I included all codes and standards required for this type of equipment?   Industry prefers working to nationally recognized codes and standards. Bidders are familiar with them and may well have had a hand in their preparation and adoption. If the writer and his or her organization use their own set of requirements where a code or standard could have been invoked, the bidders are sure to point it out and take exception. The best approach is to call out the code or standard and then note the modifications or additions to it. There is also the matter of legality. Some codes must be invoked when applicable because the law requires their use.

If I am using supporting specifications, have I referenced the correct ones? Frequently, national codes will assign the same item of equipment to one of several classes, depending upon its specific use in a system. Each class has its own unique requirements. The writer's organization may have standard supporting specifications, such as fabrication specifications, for each class. The writer must then make sure he or she has called out the correct ones for the particular equipment classification.

Have I covered all points necessary to completely define the design of this equipment? If the writer is not familiar with the equipment covered by the specification, he or she is well advised to seek advice from those who are. Otherwise, the reviewers may write the specification themselves by way of their comments. Some writers prefer this, but it is a haphazard approach. One never knows what a reviewer's current workload is, or how much time and thought he or she will devote to a particular specification.

Have I covered material processing and fabrication in sufficient detail to ensure a product of the desired quality and durability without unnecessarily restricting potential suppliers? This area is frequently the source of most bidders' exceptions to a particular specification. For most equipment the major costs are related to processing and fabrication. Bidders prefer as much leeway as possible so that they may minimize costs in a competitive situation. However, there is always the possibility that some bidders will go too far and compromise quality or durability as a result. The writer must carefully determine how far to go with details and restrictions in order to get a satisfactory product at reasonable cost.

Have I included enough tests to assure that the equipment is capable of meeting all design requirements? The testing of most standard types of equipment is covered by one or more of the national codes and standards. In many instances these tests are sufficient. However, the situation wherein equipment has been modified or will be used in a new way frequently arises. Then, the codes and standards will not be adequate, and the writer must determine and specify appropriate modifications of or additions to these documents.

Have I adequately covered all other requirements for shipping preparation and installation and operational documentation? For the writer's organization these items will probably be standard. When the writer's organization is working for some other group, these may have to conform to that group's standards. Accordingly, the writer must make the specification reflect customer requirements.

Are all requirements in the correct locations? Most problems with incorrectly located requirements arise in the areas of manufacturing and testing. The testing requirements section is for tests accomplished on the complete item only. It should not include special tests of parts or subassemblies accomplished for quality verification prior to their installation. Also, tests or examinations of basic material properties, performed before material is used, come under material procurement requirements and should not be mixed in with tests or examinations of finished parts. This is particularly true for the techniques of nondestructive examination. The writer should also make sure that the materials selection subsection does not contain requirements that belong in the material processing subsections. Specifications of heat treatment ranges and hardnesses, for instance, come under processing, not selection.

Have I complied with all requirements applicable to this type of specification for this equipment and its intended use? The writer should make sure that he or she has included any and all mandatory codes and standards. For some types of equipment, government agencies and insurance underwriters require the application of selected codes and standards. Also, the writer's organization or its customer, or both, may have generic requirements that are contractually binding.

Chapter 15

The Reviewer's Work

15.1 THE REVIEWER'S CONSIDERATIONS

The purpose of reviews, in the simplest terms, is to get the best specification possible out of the organization. A specification is expected to satisfy customer and regulatory agency requirements, to elicit adequate and comparable bids, and to result in satisfactory equipment at a reasonably low cost. To these ends, the expertise of a selected group of reviewers is utilized. Each one of them should be experienced enough in his or her area to know what a good specification should contain.

When a reviewer gets a specification, he or she has the advantage of seeing the document for the first time in its complete form. This often makes it easier to see flaws and omissions than it is for the writer. The writer is very close to the contents of the specification, having spent considerable time building it up. This long familiarity with the document often makes it difficult for the writer to see its weaknesses. The writer may assume certain things to be self-evident when they need explicit statement. It is important for the reviewer to note these items in his or her comments. The above purpose must constantly be in the reviewer's mind, and he or she should judge the technical content of the specification in view of it.

15.2 THE REVIEWER'S CHECKLIST

The following questions comprise the reviewer's checklist in appendix J. Once again the discussion will indicate the scope of each question so that the reviewer will know what to look for.

Is the coverage of requirements complete? Depending upon the area in which requirements are incomplete, the results may range from trivial to disastrous, from slightly increased power or maintenance costs to failure of the equipment to perform as required under some conditions, causing failure or the necessity of expensive and time-consuming modifications. The reviewers are expected to make sure that such problems do not arise.

Are all appropriate company documents used as references or as sources for the contents of this specification? The reviewer should know what standard company documents exist for use in his or her area. Perhaps the reviewer had a hand in their preparation. Such company standards represent the experience of the organization gathered over the years in a codified form. They should be used unless there is a good reason for not doing so in specific instances.

Are all required or appropriate codes and standards applicable to this equipment called for? The use of some codes and standards is required by government regulatory organizations. Others are widely recognized, tried and proven, and familiar to most or all suppliers. The use of recognized codes and standards is always advisable; they represent the industry consensus on how to produce satisfactory equipment. Also, suppliers usually give lower quotations when they work to familiar documents. The reviewer should be well acquainted with codes and standards applicable to his or her area.

Are there any other references that should be used? The reviewer must be aware of any special compilations of design information or materials data meant for use on the type of equipment being considered and its intended service. Many such references are concerned with long-term environmental effects, which must be taken into account when considerations of durability, reliability, and safety are important.

Is there an excess of purely descriptive material?   In general, telling the suppliers things they do not need to know is a waste of time and paper. Usually, the suppliers know much more about applications of and requirements for their type of equipment than either the specification writer or the reviewer. Their specifications analyst is only interested in the requirements he or she must identify and pass on. The reviewer should visualize himself or herself in the place of the supplier's personnel and ask whether he or she would really need to know everything that is in the specification.

Are unnecessary explanations for requirements included?   Requirements should not be explained or justified; just stated. An experienced supplier knows and understands the need for different requirements and can or cannot meet them. The supplier's bid will either accept them or take specific exceptions. Including explanations in a specification may lead to questions and arguments that waste time and money without improving the situation or changing the requirements.

Are unnecessary discussions of the origins of conditions, such as operating environments or transients, included?   The supplier needs only to know what the conditions or transients are, and not how or why they came into existence. Such information is superfluous and also a waste of time and paper. The reviewer should call all such unnecessary items to the writer's attention.

Are the requirements adequate for their intended purpose?   Getting satisfactory equipment built to an inadequate specification is highly improbable, if not impossible. Recognizing inadequate or incomplete requirements and identifying them for the writer is one of the most important parts of the reviewer's work. Here is where the reviewer's experience and effort are most valuable to the organization.

Are the requirements excessive?   This is the other side of the coin. Excessive requirements result in unnecessarily high costs and sometimes eliminate potentially good suppliers. Here, the reviewer should let the writer know where to ease up and keep the requirements within reasonable limits.

Are all special company requirements for this type of equipment included?   Organizations often have their own special requirements for certain types or applications of equipment that have not yet been placed in their own standards or reference specifications. The reviewer should be aware of all such requirements in his or her assigned area so that he or she can point out their omission to the writer.

Are the requirements clear and unambiguous?   Vague and ambiguous requirements can and often do result in large amounts of time being spent to clarify discussions with suppliers, revisions to the specification, and perhaps requests for new quotations. It is particularly frustrating and expensive to get a set of noncomparable bids, find out that the cause is one or more poorly stated requirements, and have to go out for rebids. Protecting the organization from these problems is another one of the reviewer's most important tasks.

Has the writer tried to cover a lack of specific requirements with general references to codes and standards?   Frequently, inexperienced writers will try to obtain blanket coverage of an area by writing in requirements such as, "For these structures American Welding Society standards shall apply," "These materials shall conform to the American Society for Testing and Materials Standards," or "All qualifications shall be in accordance with the Institute of Electrical and Electronic Engineers standards." Such statements, in effect, add to the specification a whole shelf of documents of which 99 percent may not be applicable. Here, the reviewer should assist the writer in selecting the correct specific standards, or in setting appropriate specific requirements.

Are any of my comments strictly editorial, and therefore within the editor's area of review?   Let the editor do the editorial work. Forcing the writer to dig through redundant editorial comments in order to find what may be only a few good noneditorial comments is wasting time. Of course, if the organization doesn't have an edior, the situation is different. The problem then is to get the editorial work done without a lot of redundancy. One solution is to circulate a copy for editorial corrections only. Another is for the reviewers to contact each other and decide how the editorial review will be handled in order to avoid duplication of effort.

Are all of my comments necessary for improving the specification, or am I simply trying to get the writer to do his or her job my way?   Many organizations require their specification writers to produce a formal, written response to all reviewers' comments. The bane of a specification writer's working life is a reviewer who buries a few good comments in a large number of others, which expound the reviewer's preferences, prejudices, and opinions while contributing nothing to improvement of the specification's technical content. Such comments waste time for both the reviewer and the writer, and generate resistance to all comments. Reviewers are well advised to confine their comments to the technically necessary, and let the writer handle the writing.

Chapter 16

The Editor's Work

16.1 EDITORIAL CONSIDERATIONS

The writer and the reviewers are solely responsible for the technical content of the specification. In an organization that has comprehensive standards for specifications, the editor is responsible for just about everything else. If such standards are lacking, the division of responsibility is less certain and may be left up to personal preference.

The editor is the guardian of the organization's professional image as it is presented by the specifications it sends out. Consequently, the editor has a much wider range of concern than the writer or the reviewer. Above all, the editor must ensure uniformity and consistency in the way an organization's specifications present their material. This goes beyond the usual, widely accepted considerations, such as when to capitalize, when to underline, and when to indent. It includes consideration of all matters of form and organization that are inherent in what is accepted as good specification practice.

16.2 THE EDITOR'S CHECKLIST

We will proceed through the questions that make up the Editor's Checklist in appendix K. The list includes some general editorial practices as well as those directly related to engineering specifications. By checking specifications against each of these points, the editor will ensure consistency and quality.

Has the typist followed the standard for specifications? Presumably, an organization has a standard for each type of document it sends out. The standards for specifications, reports, manuals, and proposals undoubtedly would be similar. Typists, however, frequently ignore the differences and type everything in the same format. The editor should watch for this, unless he or she is also in charge of the typing.

Has the correct top-level format been used? If an organization uses several formats to satisfy internal or customer-imposed requirements, the editor should verify that the correct one has been used. An inexperienced writer may have made a wrong selection.

Does the table of contents correspond to the first- and second-level section titles? Writers sometimes list every paragraph title, and they sometimes reword the titles for the table of contents. The editor should check for consistency here. (See chapter 12.)

Is the division into second, third, and fourth levels balanced? This is an editorial fine point. The editor should scan the three main areas (design, manufacturing, and testing) and note whether the presentation shifts from long final paragraphs at one level to short final paragraphs at the next lower level within an area. This variation is acceptable between areas, because some organizations consider themselves strong in the design area and flexible in the other two, while other organizations are shop-oriented and more tolerant of design variations. However, within an area such variation usually indicates the writer's uncritical acceptance of source material as written.

Is the titling of subsections and paragraphs consistent? This is another fine point. Some writers begin a title with the article "the" whenever suitable. Some writers prefer to make the subjects always singular or always plural. If the titles show a mixture of the above practices, it usually signals the writer's direct copying from material sources.

113

Have bottom-line titles been avoided? Placing a title on the bottom line of a page without at least one line of text below it is poor practice.

In parameter lists, have split compound units been avoided? Compound units, such as those for control valve leakage (cc/min/in. seat dia./lb. pressure drop), should be placed on one line, and not split between two lines.

Does the scope section contain standard informational material required by the organization? Such material, if it exists, is usually contained in an organizational standard on specification requirements. If the specification has been written for a customer, the material is usually contained in a customer supplied project guide or manual.

Does the scope section contain requirements that should appear elsewhere? If the word "shall" appears anywhere in the scope section, it indicates an incorrectly placed requirement. There may also be a correctly placed redundant statement of the requirement in one of the other sections. The editor should delete any requirements from the scope section, determine where they should be placed, and place them in their correct location.

Is the applicable documents section correctly organized? (See chapter 6.)

Is each listing complete: (See Chapter 6.)

Are all documents listed in this section called out in the requirements sections, and vice versa? The editor may have to do a considerable amount of cross-checking here. (See chapter 6.)

Does the applicable documents section contain requirements that should appear elsewhere? Once again, the appearance of the word "shall" indicates an incorrectly placed requirement. Usually, it does not occur in the listings themselves but in preceding, expanded standard statements. Frequently, it is a carryover from construction specification practice. The editor should correct this in the same way as for the scope section.

Has the limit on levels been observed in the requirements sections? This point is self-explanatory. The editor should be particularly watchful in any very long sections. (See chapter 11.)

If dependent multiple subsections at the same level have been used, does the specification move up one or more levels immediately afterward? This may seem to be a fine point of minor importance, but failure to observe it frequently leads to interpretative difficulties, especially in the materials and manufacturing requirements sections. (See chapter 11.)

Are all items in lists short enough, and do they contain one requirement only? Here the editor should make certain that the writer is not using long items in a list to avoid the limit on levels. (See chapter 11.)

Have individual requirements been separated? (See chapter 11.)

Is the ordering of requirements, modifications, and exceptions consistent? (See chapter 10.)

Have sublists been used only for breaking main list items down to their simplest components? Once again, the editor should make certain the writer is not trying to avoid the limit on levels. This is relatively easy to detect. In the extreme, a writer may take a paragraph containing several requirements, break it down into topic outline form, and change it to one or more list items with sublists. (See chapter 11.)

Have the requirements been clearly and directly stated? This is the point at which the greatest disagreements between writers and editors often arise. What is perfectly clear to the writer is vague or ambiguous to the editor. What is obviously correct to the editor is wrong to the writer in the light of his or her specialized knowledge. Carefully considering and resolving all such disagreements is well worthwhile. It almost always results in a superior specification. (See chapter 10.)

Are multiple-requirement entries correctly used? If the editor is not very familiar with the type of equipment covered by the specification, he or she may have to do some investigating here. (See chapter 11.)

Are the terms "shall" and "will" correctly used? Incorrect use of these terms in an issued specification is one of the surest signs of inept writing and editing, even though it usually does no harm. (See chapter 9.)

Are all cross-references correctly numbered? Incorrectly numbered cross-references in a long specification can be a source of much frustration to the specification's users. Usually, they result from extensive revisions to the review draft. The editor should check the numbering of cross-references and consult the writer on all points of uncertainty.

Do subsection titles adequately describe the contents? By checking this point carefully the editor may find he or she can improve the specification by either increasing or decreasing (and rearranging) the number of subsections.

Do any multiple subsections follow a logical order? The editor can check this point while he or she checks the preceding one. Once again, a better arrangement may be possible.

Have requirements that belong in other documents of the procurement package been included? Frequently, inexperienced writers will put contractual (legal) requirements in the technical specification. These should be removed and placed in the contract proper. Many organizations require contractors to submit schedules, reports, shipping information, and so on. Any such requirements that are not directly related to the hardware should not be in the technical specification. The editor should be familiar enough with the organization's procurement documentation to know where such requirements should go.

Have all grammatical errors been corrected? To an editor, this point is self-explanatory.

Have all administrative requirements of the organization been met? This gets down to the purely clerical part of the editor's task: checking to see if all administrative control pages have been included, if the number has been assigned correctly, if the title is categorized correctly, if it has been assigned correctly to a specification file, if the distribution list is correct, if the required signatures have been obtained, and so forth.

Chapter 17

The First Procurement

17.1 INITIAL QUOTATIONS

Staffs of engineering and purchasing departments often hold the opinion that no specification is really finished until it has gone through one procurement cycle and has been placed on contract. This gives a group of potential suppliers, faced with the prospect of working to the specification, the opportunity for reviewing and commenting upon it. In chapter 14 it was pointed out that the writer could seek the advice of potential suppliers during preparation of the specification. All too often such requests are given polite but cursory treatment. Many suppliers consider a response as "free engineering," which must be charged to their general overhead. However, suppliers do have budgets for responding to formal requests for quotations, and they prefer making their considered comments within that context. Thus, the most significant suppliers' comments will accompany their initial quotations.

Figure 17.1 shows a brief outline of the complete process, from writing through procurement, wherein a specification reaches its finished form as a contractual document. Chapter 13 contained a

117

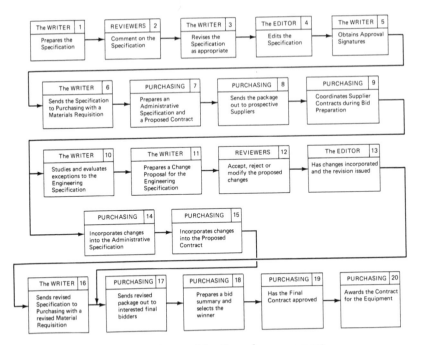

Figure 17.1 From writing through procurement

description of the process leading up to the initial issue of a specification. This is summarized in the first five blocks at the top of Figure 17.1. In the next step, as shown in block 6, the writer prepares a materials requisition and forwards it to purchasing along with the specification (presuming the writer has continuing responsibilities). Purchasing then prepares an administrative specification and a proposed contract (presuming a standard purchase order will not be used) as discussed in chapter 1, and sends them out along with the engineering specification in a request for quotations. Or, a preliminary purchase order may be used as the vehicle for obtaining quotations. These steps are shown in blocks 7 and 8. Purchasing then coordinates all contacts with suppliers during their bid preparation, as shown in block 9. Comments on and exceptions to the engineering specification will be sent to the writer. Those pertaining to the administrative specification and the contract will be handled by purchasing. If the differences are massive, contradictory, or both, a bidders' conference may be necessary. If not, they can be dealt with by letter. Under the provisions of the Uniform Commercial Code, any clarification or explanation sent to one bidder must be sent to all, even though they asked for none. This keeps anyone from gaining an unfair advantage. This, then, is the point where most undiscovered weaknesses and faults in the procurement documentation become evident.

17.2 REVISING THE DOCUMENTS

When all exceptions and comments are in and all clarifications and explanations sent out, the writer has to evaluate all of the material and determine the appropriate revisions. Block 10 of Figure 17.1 illustrates this point. If a number of bidders take approximately the same exception at a particular point and recommend similar changes, they are probably right. Usually, this is no cause for concern. The problems arise with lone exceptions. Perhaps the bidder is simply trying to protect or enhance his or her competitive position. Or perhaps the bidder knows something his or her competitors and the writer do not know. No exceptions can be ignored; in the end, a specification writer must consider them all.

When this task is complete the writer prepares a change proposal, delineating the revisions he or she thinks should be made, and sends it to the reviewers. This is shown in block 12 of the figure. Concurrent with the above effort, purchasing must revise its documents as shown in blocks 11 and 13. Some of the requirements in the administrative specification may have been incompatible with the operations of one or more bidders. Some of the terms and conditions in the contract (or preliminary purchase order) may have been impractical or unacceptable to some of the bidders. Thus, purchasing has a parallel document revision task. The specification writer handles the engineering specification, and purchasing handles the other documents.

When the reviewers have studied the change proposal, they may advise acceptance, rejection, or modification of the proposed revisions, as shown in block 14. This is the point at which differences between specialized experts are most likely to arise. Each will marshal his or her expertise and experience in support of his or her position. The writer must listen closely, proceed carefully, and, above all, be tactful. When this stage is complete, the writer sends the agreed-upon revisions to the editor, who has them incorporated into the specification, and the revision is issued according to standard procedures. Block 15 shows this culmination of the revision process.

17.3 FINAL QUOTATIONS

One task remains for the writer. He or she must send the revised specification to purchasing along with a revised material requisition, as shown in block 16. Purchasing then sends all of the revised documentation out to remaining bidders for the "best and final" bids. If the job has been done well, the bids will come in without exceptions. The bidders will know just what is expected of them. Purchasing then prepares a summary of the bids and selects the winning bidder. The final contract for the specified equipment is approved and awarded to the winning bidder. These final steps are shown in blocks 17 through 20. During this last phase the writer's function is that of a spectator.

The procurement process has been generalized, and perhaps idealized in this chapter. Each organization will have its own variations. This chapter has attempted to present the essence of the process, showing the interplay among the writer, the reviewers, the editor, and the purchasing staff. Things can always go wrong at any point. However, if everyone involved understands the essence of the process in which they are taking part, the probability of difficulty is considerably lessened.

Glossary

This glossary contains definitions for the following common terms, as used in this book:

1) System
2) Subsystem
3) Component
4) Assembly
5) Part
6) Material
7) Noninteractive subsystems and components
8) Interactive subsystems and components
9) Passive components and parts
10) Active components and parts
11) Passive Subsystems
12) Active subsystems
13) Reliability
14) Durability

15) Maintainability

16) Interchangeability

Frequently, these terms are used in a vague and overlapping way. The definitions given here are most useful for describing specification-writing concerns. The structure of specification hierarchies and arrangements is dependent upon them. A knowledge of these definitions is necessary for a clear and accurate understanding of much of the material presented in this book.

The first six are related to the content of Figure 2.1 and its related discussion (see page 8).

1. *System*. A system is a group of interconnected or interrelated components that can accomplish a variety of functions beyond the capabilities of its individual components acting by themselves. An automobile is an example of a personnel transport system, a microwave oven is a food processing system, and a room air conditioner is an environmental modification system. All accomplish functions beyond the capabilities of their components (their individual pieces).

2. *Subsystem*. A subsystem is one of several segments of a system, capable of accomplishing a select group of the functions of the complete system, or concurrent intermediate functions, that are beyond the capabilities of the individual components of the subsystem. Frequently, subsystems can function independently and are treated as smaller-scale systems. For this reason the major subsystems of something complex are often called its systems, and the name of the item does not include the term "system" (automobile, microwave oven, air conditioner). The engine, transmission, and running gear of an automobile form its propulsion subsystem or system, depending upon how the terms are used. The fuel tank, pump, connecting lines, and filter form its fuel supply subsystem or system, performing the function of supplying fuel to the running engine. An air conditioner built into a heating, ventilating, and cooling unit is a subsystem or system performing the cooling function. A television set may be considered a subsystem or system of a home entertainment center (along with record and tape players and game-playing attachments), or the reception subsystem or system of an information transmission medium. Thus, the same group of components, connected in the same way, may be called a system or a subsystem and may perform the same function for more than one overall group of functions.

3. *Component*. A component is an item within a system or subsystem that performs a specialized system or subsystem function, or a limited intermediate function. The limited intermediate function

is one necessary for the performance of final or end functions of the system or subsystem. An example is a pump used to fill an elevated tank for a water distribution system. Others are a bridge rectifier converting AC to DC for a power supply, a transfer gearbox for changing shaft speeds, a starter for electric motors, and an electrical power distribution panel. In an automobile, the fuel pump, carburetor, starter, and generator are components.

Components may be very complex and divisible into other components performing other intermediate functions related to the overall component function. A valve for fluid flow control with a pneumatic, hydraulic, or electric actuator for positioning the valve is an example. Another is a motor-generator set, where a motor operating on 60-cycle current runs a generator capable of producing a different frequency current. Thus, complex components consist of an assembly of other components and parts, and may be treated in a manner similar to systems.

4. *Assembly.* An assembly is a component or group of components and parts held together by various nonpermanent attachment methods (such as bolts and nuts, threaded connections, or clamps) that can be disassembled without the use of destructive methods. It can then be reassembled. The terms "assembly" and "component" are often used interchangeably. Most often this term is applied to a component by the organization putting it together as a new item. A small cabinet for electronic equipment, consisting of a chassis, a front panel, and a case held together by bolts and nuts, is an assembly. So is an automobile door when the window glass, lifting mechanism, lock, handle, trim, and upholstery have been added.

The term "assembly" is almost universally used on drawings that show how components and part are put together to make higher-level assemblies. Often, this indicates completeness for some purpose. If what is called an automobile engine is the combination of block and head assemblies, then the engine assembly consists of that plus the carburetor, starter, generator, fuel pump, distributor, and so on, ready for installation in the automobile.

5. *Part.* A part is an item put together by some permanent manufacturing process, using other parts, that cannot be disassembled without the use of destructive methods. A mounting stand welded together from simple structural steel shapes, no matter what its size, is a part. An automobile door weldment is the basic part of the door assembly. A wiring harness with soldered or crimped terminals and a bonded covering is also a part. A printed circuit board with all of its parts soldered in place is yet another part, so complex that many may consider it an integrated component or subsystem.

A part is also an item, made from stock material, ready for use immediately following its manufacturing process. Bolts, nuts, terminals, and insulators are parts and are sometimes called *piece parts* in view of their simplicity.

Some manufacturers will take what could be an assembly and enclose it in a case so that the casing joint is permanent and has to be cut open destructively before any further disassembly can take place. This creates a larger, more complex replaceable part. Examples are automobile voltage regulators, numerous small electric motors, and small gear sets for light mechanisms.

6. *Material.* A material is something that cannot be used as a part without first being subjected to some manufacturing process. Frequently, a distinction is made between two types of materials: raw and finished. A manufacturer who makes bolts and nuts refers to the bar stock as raw material (or stock) and the items are finished parts. A manufacturer who buys bolts and nuts for use in assembled items calls them finished materials. A manufacturer who makes shielded and insulated coaxial cable calls the wire, plastic, bonding material, etc., raw materials, and the reel of cable his finished product. A manufacturer who buys the reel of cable and makes leads by cutting it to length, stripping the ends, and attaching terminals, calls the cable raw material and the leads finished materials for use in assemblies along with the bolts and nuts. Thus, the terms "finished materials" and "finished parts" are used in a relative manner, which may be misleading. However, the term "raw material" (or "raw stock") always implies that a manufacturing process has to take place prior to use in an assembly.

The term "parts list" and "materials list" are often used interchangeably, although there is a difference. In a strict sense, a parts list would call out finished parts and a materials list would call out raw materials. Many such lists are actually combined parts and materials lists.

The next two definitions are related to the contents of Figures 3.1 through 3.4 and their related discussion.

7. *Noninteractive Subsystems and Components.* Noninteractive subsystems and components are those whose performance (function) is independent of the performance of related subsystems and components within the same system, subsystem, or assembly of components. The radio in an automobile is independent of the engine, the air conditioner, and the suspension. The timer on a microwave oven is independent of the radiation generation subsystem. It simply determines the period of operation, independent of the effectiveness of operation.

The air-cooling subsystem of a heating, ventilating, and air-conditioning system is independent of the operation of the burners. A specified set of temperature limits determines which one is operative.

8. *Interactive Subsystems and Components.* Interactive subsystems and components are those whose performance (function) is dependent upon the performance of other subsystems and components within the same system, subsystem, or assembly of components. The performance of an automobile engine is dependent upon the performance of its fuel supply and ignition subsystems. The performance of its suspension is dependent upon the interaction of springs, shock absorbers, and other parts in the linkage. The performance of a television set depends upon the correct functioning and interaction of the antenna, a number of electronic subsystems, and the picture tube and speaker.

The requirements for one interactive item cannot be completely determined without determining at least some of those for all other interacting items. This means that a model for the group of interacting items must be set up with all of the interactions included. No such model is necessary for noninteractive items. This also means that the higher-level system or subsystem specification, while not completely defining all of the requirements for its components, must provide the basis for setting up whatever models are necessary.

The last eight definitions are related to the contents of Figure 14.2 (p. 104) and its related discussion. The first four concepts are less well understood than any of the others, thus requiring more extensive discussion.

9. *Passive Components and Parts.* A passive component or part is one whose function is completely determined by a fixed geometry and its materials of construction. The parameter or parameters determining its function are constant and independent of its installation. Resistors, capacitors, inductors, and transformers are passive electrical parts. Their resistance, capacitance, inductance, and voltage ratios are fixed by design and do not vary along with other circuit parameters. Tanks, orifices, nozzles, and heat exchangers are passive mechanical components and parts. The volume of a tank is fixed by design as is the capacity of a capacitor. Generalized flow and heat transfer coefficients for orifices, nozzles, and heat exchangers are fixed by design and do not change along with other installation conditions. Connecting cables and wires in electric and electronic circuits, and piping in fluid systems are passive components. All fixed supporting structures are passive.

If the defining parameter of an item is made variable by means of geometric (positional) adjustments, with the variation accomplished by direct, immediate manual operation, and if the function of the item does not depend upon the motion but only upon its end points, that item is still considered passive. Thus, variable resistors, capacitors, inductors, and transformers, wherein the setting operation is accomplished by hand, are passive. A valve that must be operated by a handle or handwheel, and an adjustable height stand operated by some form of jack are passive mechanisms. They are mechanisms because relative motion between their parts is possible, and they are passive because it is the end points or positions that count, not the intermediate motion.

10. *Active Components and Parts.* An active component or part is one whose function is determined by relative motion of parts, the conversion or modulation of energy (operational heat loss not included), or both. The parameter or parameters determining its function are variable, and particular levels are dependent upon its installation. That is, its function is defined by a performance curve or a family of performance curves, and the conditions of its installation determine an operating point on those curves. Transistors and vacuum tubes (triodes and up) are active electronic parts. Their operating characteristics are defined by sets of installation-dependent curves defining energy modulation. Batteries are active; they convert chemical to electrical energy through the medium of an electrolyte. Pumps and turbines are active mechanical components; they have moving parts and convert mechanical energy from one form to another. Engines are active components; they have moving parts and perform conversions between chemical and mechanical energy through the medium of heated gases. Transmissions are active; they have moving parts performing a motion-dependent function. Electric motors and generators are active electromechanical components; they have moving parts and perform conversions between electrical and mechanical energy.

The function of active mechanical components, then, always involves mechanical motion and may also involve energy conversion. The function of active electrical components always involves energy modulation and may also involve energy conversion. The functions of active electromechanical components always involve both mechanical motion and energy conversion. A two-position solenoid-operated device, such as a switch, is considered active. Its motion has the positional characteristic of a passive device, but the electromechanical energy conversion involved classifies it as active.

Whenever the functions of a component, due to its design, are self initiated, initiated by the action of other components, or initiated by indirect manual action, such as closing a switch or setting a dial, that item is considered active. A diode is considered an active part because its function of changing current flow conditions is self initiated due to its design. A logic gate is considered active because its function is initiated by its design and the action of other components or parts. A check valve is considered active because it has a moving part (positionally functional) which responds to its internal flow conditions. A hand operated pump is considered active because its function is dependent upon continuous motion, and thus upon continuous manual operation, which involves the conversion or transfer of energy from the operator to the mechanism. Position alone is not a determinant.

CONVERTING PASSIVE ITEMS TO ACTIVE

Passive components or parts may be made active by the addition of an actuating mechanism. A passive manual valve may be made active by the addition of an electric, hydraulic, or pneumatic actuator. A passive variable resistor, capacitor, inductor, or transformer may be made active by the addition of an electric motor drive. Thus, the addition by assembly of an active component to a passive component makes a more complex active component. By now a fundamental point should be clear. An item that meets only passive criteria is considered passive. An item that meets any one of the active criteria is considered active regardless of the passive criteria it also satisfies.

As complex active assemblies are broken down into simpler components and parts, the ratio of passive to active items increases. Indeed, some active items, like pumps and electric motors, can be broken down into nothing but passive parts. How the parts are put together and used makes the assembly active.

11. *Passive Subsystems.* A passive subsystem is one that produces its desired functions with its intermediate operations under direct, concurrent, or sequential manipulation by a human operator, or by a manual or automatic control system. By itself, a passive subsystem can operate only in a steady or equilibrium state. It cannot start itself, stop itself, or change to a new set of equilibrium conditions. Passive subsystems may contain passive components and parts only, or both active and passive components and parts. If they contain passive items only, direct control by a human operator is necessary. If control is accomplished indirectly or automatically, at least one active item is necessary. Active components and parts that directly affect operation of the passive subsystem are called its final control elements. Since the passive subsystem cannot function without them, they are considered components and parts of that subsystem.

12. *Active Subsystems.* An active subsystem is one that controls the operation (functioning) of a passive subsystem or component, so that it produces the desired end result without direct, concurrent, or sequential manipulation by a human operator. Active subsystems are most frequently called controls or control systems. If an active subsystem requires indirect or remote concurrent manipulation by a human operator, it is called a manual control or manual control system. If it does not require a human operator it is called an automatic control or automatic control system. An active subsystem (or active part of a system) starts a passive subsystem (or passive part of a system), stops it, and alters its total state of equilibrium by producing local sets of nonequilibrium or transient conditions. It can also maintain equilibrium of the passive subsystem under the action of externally caused disturbances. Active subsystems contain both active and passive components and parts. Final control elements in the related passive subsystem are at the terminal points of the active subsystem. Other components and parts mounted in the passive subsystem for monitoring its state may be at additional terminal points of the active subsystem. These items, called instruments or sensors, are considered part of the active subsystem. Functioning of the passive subsystem does not depend on their operation to produce its own output, although it may not operate very well or for more than a short time without them and the active subsystem.

Examples

A host of examples for illustrating the passive and active can be found in the ubiquitous automobile. The fuel supply subsystem consisting of a tank, pump, connecting lines, and filter is a passive subsystem with one active component: the pump. The fuel pump is run by the engine or an electric motor and is operated automatically by the engine in the first instance, and by indirect manual switching in the second. For diesel automobiles one can get an auxiliary tank with a branch line and a valve connected to the main fuel line. If the valve is manually operated the auxiliary subsystem is passive, with all passive components. If the valve is solenoid-operated, with a switch on the dashboard, it is another passive subsystem with one active component, for indirect manual operation. If the fuel level indicator in the main tank is connected to the solenoid valve and battery so that when the main tank level falls below a fixed point the valve opens, it is a simple active subsystem automatically controlling the passive auxiliary fuel subsystem. The solenoid valve is the final control element, the fuel level indicator is the instrument (or sensor), and the switching device activating the solenoid in response to the main fuel tank level signal is the control.

The hand crank, linkage, cabling, and window form a simple, passive, positionable subsystem for varying the size of an opening in the door. If the hand crank is replaced with an electric motor drive and a switch, an active component has been added but the window subsystem is still passive, controlled by a manual active subsystem consisting of the switch, wiring, and battery. If a moisture detector is added and connected so that rain causes the motor to start and close the window, there is another pair of subsystems, one passive and one active and automatic.

These examples illustrate some fundamental points. Both subsystem pairs (fuel and window) require the battery (or generator) as an energy supply for their active subsystems. Every active subsystem has at least one active component or part, and if there is only one it is the energy source for subsystem operation. Such a source is required for every active subsystem. Also, a component or part may be common to more than one subsystem. An automobile battery or generator is an active item common to several subsystems. The ignition switch is a passive item common to several subsystems.

A basic carburetor is a complex passive component whose function is dependent on the position of its adjustments and variable features. The accelerator pedal, linkage, and throttle plate form a manual speed-setting subsystem (passive) consisting of all passive components and parts. How the speed changes between settings depends upon the design of all the various parts, the settings of the adjustable features, and how fast the operator chooses to move the pedal. If a cruise control feature is added, sensing speed and repositioning the linkage and throttle plate as necessary with a servomotor, a more complex active and automatic subsystem has been created with both active and passive components and parts. The radiator, fan, connecting hoses, water jacket, and water pump form a passive heat removal subsystem with two active components: the fan and the pump. These are run automatically by the engine. Of course, the fan may be operated by an electric motor, which is started by a temperature-activated switch on the water jacket. Powered by the battery, this arrangement can continue cooling operation after the engine has been shut down. In this instance one of the active components, the fan, has been made more complex in order to provide more flexible operation.

The suspension subsystem of an automobile is passive. However, its primary components and parts, the shock absorbers and springs, are active. Their function depends upon relative motion within themselves, determined by relative motion between the body and the wheels and axles. The function of a shock absorber always depends upon motion of parts, so it is always active. A spring, however, can be considered active or passive depending upon its use. If its use involves motion, as in an automobile suspension when the machine is moving, it is active. If it is used to provide a load, or to support a load in a fixed position as in an automobile when the machine is standing still, it is passive. The simple spring is an interesting item when the distinction between active and passive parts is made.

The electric supply subsystem has three active components—the generator, the battery, and the voltage regulator—and several passive parts, including cables and wiring. The ignition subsystem has one active component—the distributor—and several passive parts, including spark plugs, the coil, and wiring. Both of these subsystems are run automatically by the engine.

The automobile radio is a complex of electronic and electromechanical subsystems, components, and parts. Its overall operation is manual, with pushbuttons and dials. It may have a station scanning device which, once started, works automatically. It may have a tape-playing subsystem with manual or automatic rewind features. Once again, the amount of activity and automation depends upon the items added and how they are used.

FINAL COMMENTS

Some systems and subsystems are compact, with active components directly attached to one another. Others may be dispersed, with passive connectors of varying lengths between active components. This fact is the source of much of the loose and sometimes confusing use of terms. A basic gas turbine has three active components—a compressor, a combustor, and a turbine—directly attached to each other. A motor-generator set also consists of two or three active components directly attached to each other. In both of these items the active components are also interactive. Such items are usually looked upon as systems or subsystems by their manufacturers, who must analyze them as such. Their users frequently look upon them as components.

In many hydraulic, pneumatic, and electric installations, active components are located in different places and are connected together by passive items such as piping and wiring. These installations are usually called systems or subsystems by both assemblers and users when they are large. When they are small and can be packaged and transported like components, they are often called components.

In the realm of electronic systems there are many which are like dispersed systems operationally, and like compact systems structurally (in the way they are assembled). Industrial examples are racks of individual items of electronic test equipment connected with cables; commercial examples are recorded music systems, likewise consisting of separate components (or subsystems), connected with cables in a common rack or cabinet.

By definition, passive items are functionally noninteractive. However, they may react to connected items in a nonfunctional but unavoidable way. Structural reactions, not related to design functions but unavoidably determined by structural designs that produce those function, fall into this area. Loads on supports and the effects of generated heat are the commonest examples.

Active items may be functionally interactive or noninteractive. They may interact directly or through passive connections. They also react to connected active and passive items in a nonfunctional and unavoidable way, determined by design. Another way of viewing this situation is to consider all interactions as functions and divide them into intended (designed to produce) functions and unintended but unavoidable (designed to withstand) functions.

A question may have formed in the reader's mind by now. Is the division into two types of components, active and passive, oversimplified? A liquid storage tank is considered passive whether or not the liquid level is changing. The same consideration applies to an electrical capacitor and its charge. A simple spring is considered passive if it is used to hold position, and active if its structure is used to control motion. An electric motor transforms electrical energy into shaft motion. A pump transforms shaft motion into fluid flow. Other components perform both transformation and control functions. Perhaps a four-way division into passive, structurally active, transformationally active, and multiply active components would be more useful. Or should it be totally passive, conditionally passive, simply active, and multiply active components? Such considerations are valid and could lead to clearer and more precise definitions of requirements, particularly in the area of equipment qualification for major projects. Unfortunately, they are not in general use.

13. *Reliability.* Concern for reliability (occasionally called dependability) may be understood in the context of the following question: Will this item function as intended, when intended, and as often as intended? If the answer is yes, the item is considered reliable. Primarily, then, reliability is related to on/off cyclic operation and its repeatability under various conditions.

14. *Durability.* Concern for durability may be understood in the context of the following question: Will this item function as intended,

for as long as intended, regardless of how long it was previously on standby (not functioning)? If the answer is yes, the item is considered durable. Primarily, then, durability is related to resistance to aging and wear under various conditions.

THE RELATIONSHIP BETWEEN RELIABILITY AND DURABILITY

Reliability and durability are closely related, and much of the time it is difficult or impossible to treat them separately. Also, desired reliability is most closely related to operational requirements, and desired durability to structural requirements. An equipment item has a defined operating life, and within that life it may have long periods on standby, long periods of continuous operation (steady or cyclic), and periods of intermittent operation. The variety is infinite. Thus, the variation in emphasis placed on reliability and durability is considerable.

15. *Maintainability.* An item is considered maintainable if it can be serviced (preventive maintenance) or repaired (corrective maintenance) for considerably less cost than would be required for manufacture of a new item. Primarily, then, maintainability is related to how an item is designed and assembled. Ideally, all maintenance should be performable with the item left installed. In such instances maintainability implies accessibility. If the ideal situation is not attainable, the next choice may be to minimize shutdown time by having the item replaceable with a new or rebuilt spare. In these instances maintainability implies simplicity of installation, adjustment, and nonpermanent attachment. Occasionally, an item is designed for serviceability but not repairability. Some small electric motors are constructed so that their bearings can be oiled but not replaced if they burn out from lack of oil.

16. *Interchangeability.* An item is considered interchangeable if it can replace another nonidentical item in an installation without modifications being required and if it can perform the same function after installation as the replaced item. Interchangeability, then, implies that certain critical functions and physical characteristics are identical. This is not the same as commonality, which means that an item has several different applications, and spares kept for one of them can be used for the others. Commonality is sometimes called interchangeablity, and this can cause confusion.

Maintainability and interchangeability are determined by requirements for design features, as are two other "ilities," somewhat similar, and of much more concern to the military than to commercial operations

These are transportability and operability. More than anyone, the military has equipment which must be fixed in place for use, and also disassembled, moved and reassembled at other locations. Large field radar installations are an example, as are field repair facilities for military equipment. Commercial examples of "traveling equipment" are portable public address systems and carnival rides. Since military equipment frequently must be operated under adverse conditions, and by a wide variety of human operators, human capabilities must be considered. This leads into what is known as human engineering and man/machine interfaces; an area in which the military has long been the acknowledged leader.

Appendixes

APPENDIX A:
SPECIFICATION-RELATED DRAWINGS (See Chapter 3)

The following six common types of drawings are all related to components (and their specifications), are all similar in a number of respects, and are sometimes interchangeable:

1. Purchased part drawings
2. Specification control drawings
3. Equipment requirement drawings
4. Installation drawings
5. Outline drawings
6. Interface control drawings

The primary purpose of the first three is to transmit information from the purchaser to the supplier. The primary purpose of the last three is to transmit information from the supplier to the purchaser or the user.

The first three determine component characteristics. The last three are determined by supplier response to specification requirements.

1. *Purchased part drawings.* These drawings are most frequently used to buy small items in quantity. The simplest are for bolts, nuts, washers, and so on. They may also be used for more complex standard components which suppliers of a particular kind of equipment do not want to make for themselves. Consider, for instance, suppliers of hydraulic equipment. A complex piece of equipment may include electric switches, solenoids, motors, controls, sensors, wiring harnesses, and so on. The suppliers would buy these with purchased part drawings and make the hydromechanical components themselves.

Then consider suppliers of electronic test equipment. This equipment comes as systems or subsystems in movable cabinets. The suppliers would buy the cabinets, mounting racks, connector panels, cooling fans, and so on, with purchased part drawings and make the amplifiers, power supplies, signal conditioners, analyzers, and so on, themselves.

The purchased part drawing is meant to be complete in itself. It does not reference any specifications or other documents. It contains all necessary requirements to ensure installation and use of the equipment in the desired manner. Usually, it also contains a list of qualified suppliers and a list of their equipment catalogue numbers, if these are available. This enables the purchasing department to buy the items at the lowest current cost with a minimum of paperwork.

Purchased part drawings are seldom referenced in specifications sent out to other suppliers. One circumstance where this may be done is when an organization is buying more equipment than usual, and in the form of assemblies. The organization may wish to make the installation and mounting hardware on this equipment interchangeable with its own. The equipment purchase specification will then reference the appropriate purchased part drawings to guarantee this interchangeability. In this instance the specification would be the higher level, or controlling document.

2. *Specification control drawings.* These drawings always reference and are accompanied by specifications. Their basic purpose is to assign different specifications to different groups of similar components. The differences may arise from operating conditions, quality assurance requirements, code classifications, auxiliary equipment, and so forth. In many instances they replace a number of simple data sheets with a single drawing.

In addition to assigning specifications, these drawings must contain enough data, or requirements for data in the supplier's pro-proposal, to ensure installation and use of the equipment in the desired manner. To do this they must include the same informational content as a purchased part drawing. They may modify the contents of the specifications for certain items. These drawings are usually used for standard components more complex than those covered by purchased part drawings. Specification control drawings are the higher level, or controlling documents.

3. *Equipment requirement drawings.* These drawings are primarily used to buy designed-to-order equipment. They must contain all necessary information (size, shape, mountings, connections, and so on) to ensure the desired installation and use. If the purchaser wishes to impose certain configuration requirements on the equipment internals, these details may also be placed on the drawing. These internal requirements make an equipment requirement drawing substantially different from the remaining three types. Equipment requirement drawings are referenced by the specifications for the equipment covered. The specifications, then, are the controlling documents.

4. *Installation drawings.* These drawings are made by equipment suppliers to cover their standard components. The drawings contain all necessary information to ensure proper installation, including all mounting and connection details and locations. When a purchaser is buying standard equipment to a specification and is not sure just what the supplier will propose, he or she will request installation drawings along with the proposal. The purchaser will do this to make certain that the proposed equipment will fit into his or her system.

5. *Outline drawings.* These drawings are basically preliminary installation drawings for nonstandard, built-to-order equipment. The purchaser requires them from the supplier so that he or she can finalize equipment locations, arrangements, and room or cell sizes in buildings. In their preliminary form the drawings must show size, shape, mounting points, and fluid connections. In the early stages, piping connections for fluids are of more concern than electrical connections. Pipe routing is more complex and more costly than cable routing. Two right-angle bends in an electrical cable will not affect its functioning. Two such elbows in a pipe run can create bothersome design problems.

6. *Interface control drawings.* When a large project is under way involving several contractors designing different systems, each of whom will utilize a number of equipment suppliers, steps must be taken to ensure that everything fits together at the site of final assembly. The primary tool (often the only tool) for accomplishing this is the interface control drawing. The engineer or engineering group responsible for a particular piece of equipment must have the drawing prepared

and sent to all project participants who have an associated interface (must provide the mating parts).

Ideally, an interface control drawing should be prepared from a supplier's installation or outline drawing. The main reason for not using these drawings directly is that most large projects require a standard format for project documents, including interface control drawings. Also, preliminary interface control drawings are often required in advance of the selection of suppliers so that preliminary building and equipment layouts can be made.

In essence, this drawing is an equipment requirement drawing without the internal details. For this reason preliminary interface control drawings are sometimes referenced in component specifications instead of equipment requirement drawings. In this circumstance some confusion may arise as to which is the controlling document. There is a point at which all interface control drawings for a project must be finalized. The contractors must stop accommodating each other's changes. At this point an interface control drawing referenced in a component specification must be the controlling document. This can create a problem with the way the two documents reference each other.

Referencing interface control drawings in specifications meets strong resistance in some organizations. The root of the problem is the fact that these drawings are simultaneous answers to questions about "what" and "how." They are both design requirements and design solutions, and are unique in this respect. Some thought will reveal that this must be so; otherwise there can be no guarantee that components can be assembled together. In fluid systems one cannot directly attach a bolted flange on one component to a butt welding nozzle on another. Nor can one directly attach a group of ring- or tongue-type terminals to a screw-on pin-type connector for electrical components. At interfaces, design configurations meeting design requirements must be mutually compatible on both sides, arguments concerning the primacy of requirements over solutions notwithstanding.

The above discussion has presented the essence of the six drawing types. MIL-STD-100 on Engineering Drawing Practices contains a much longer list of drawing types. In most instances these represent more detailed subtypes of the above, each meant to deal with a particular procurement situation. In industry there are other less common types (some of them hybrids) derived from the older miliary types. These too were meant to deal with particular situations as they arose. An understanding of the six types described above should enable one to discern the intent and place of most, if not all, specification- and procurement-related drawings in current use.

APPENDIX B:
COMPONENT SPECIFICATION—STANDARD FORM (See Chapter 6)

1. Scope

   Topics: a. Description of the equipment covered

              b. Limitations on applications of the specification

              c. Extent of work covered

              d. Codes and standards with which the specification conforms

2. Applicable Documents

   2.1 Standard statements concerning applicability

   2.2 Company documents

   List specifications and drawings, or drawings and specifications, depending upon their order of precedence.

   2.3 Other documents

   2.4 Codes and standards

   List in order of precedence, if such an order exists.

3. Design Requirements

   3.1 Operational design criteria

   Topics: a. Applicable specifications, codes, and standards

              b. Normal operating parameters

              c. Normal operating conditions

              d. Abnormal operating parameters

              e. Abnormal operating conditions

              f. Adjustable and nonadjustable set points and limits

              g. Multiple inputs and outputs

   3.2 Structural design criteria

   Topics: a. Applicable specifications, codes, and standards

              b. Steady-state loadings (normal and abnormal)

              c. Transient loadings (normal and abnormal)

              d. Expected loading combinations

              e. Analytical considerations

   3.3 Design features

   Topics: a. Structural details

              b. Mechanical details

              c. Electrical details

   3.4 Materials selection

   Topics: a. Part groups—parts may be divided into groups on the basis of several alternatives. Some alternatives are:

                 1. Criticality and safety—parts may be grouped as critical to operation and safety, critical to operation only, and noncritical.

                 2. Material type—parts may be grouped as metallic, nonmetallic, and composite.

                 3. Functionality—parts may be grouped as stationary,

moving (rotating or reciprocating), mechanical, and electrical.

4. Durability—parts may be divided into lifetime parts and those replaceable on a scheduled or as-required basis.

5. A suitable combination of the above.

  b. Specific materials—for individual parts or groups of parts, materials may be called out as:

1. Required
2. Preferred
3. Allowable (alternatives)
4. Disallowed

3.5 Design documentation

Topics: a. Report(s)—these may cover operating and structural characteristics.

  b. Drawings—these may be the proposal drawing(s), the initial (outline and connection details) drawing(s), and the final assembly drawing(s).

4. Manufacturing Requirements

4.1 Materials procurement

Topics: a. Applicable specifications, codes, and standards
  b. Tests of material properties
  c. Materials examination and acceptance criteria
  d. Certified materials test reports
  e. Certificates of conformance (compliance)

4.2 Processes

Topics: a. Applicable specifications, codes, and standards
  b. Allowable processes and modifications or exceptions
  c. Limitations on allowable processes
  d. Disallowed processes
  e. Examination of finished parts and acceptance criteria
  f. Special dimensional measurements

4.3 Identification

Topics: a. Nameplate requirements
  b. Warning tags
  c. Other markings

4.4 Pretest cleaning

Topics: a. Cleaning materials
  b. Methods

5. Testing Requirements

5.1 Prototype (qualification) testing

Topics: a. Applicable specifications, codes, and standards
  b. Structural tests
  c. Operational (performance) tests

        d. Environmental (life) tests

        e. Delivery (disposition) of the test unit(s)

  5.2 Acceptance (production) testing

     Topics: a. Applicable specifications, codes, and standards

             b. Assembly verification tests

             c. Proof tests

             d. Operational (performance) tests

  5.3 Test documentation

     Topics: a. Test plans

             b. Instrumentation (for equipment tested and facility)

             c. Drawings (test set-up, fixtures, etc.)

             d. Procedures (pretest checks, calibrations, precautions, etc.)

             e. Form and organization of test results

             f. Verification documentation

6. Other Requirements

  6.1 Preparation for shipment

     Topics: a. Applicable specifications, codes, and standards

             b. Final cleaning

             c. Use of preservatives

             d. Packing

             e. Labeling

  6.2 Installation documentation

     Topics: a. Receiving inspection instructions

             b. Handling instructions

             c. Installation instructions

             d. Checkout procedures

  6.3 Operational documentation

     Topics: a. Operation manual(s)

             b. Maintenance manual(s)

             c. Special tools and spare parts list

APPENDIX C:
COMPONENT SPECIFICATION—MODIFIED STANDARD FORM
(See Chapter 6)

1. Scope
   See standard form outline, section 1.

2. Applicable Documents
   See standard form outline, section 2.

3. Design Requirements
   See standard form outline, subsections 3.1, 3.2, 3.3, and 3.5.

4. Material Requirements
   See standard form outline, subsections 3.4 and 4.1.

5. Fabrication Requirements
   See standard form outline, subsections 4.2, 4.3, and 4.4.

6. Testing Requirements
   See standard form outline, section 5.

7. Packing and Shipping Requirements
   See standard form outline, subsection 6.1.

8. Other Requirements
   See standard form outline, subsections 6.2 and 6.3.

APPENDIX D:
COMPONENT SPECIFICATION—MILITARY FORM (See Chapter 6)

1. Scope
   See standard form outline, section 1.

2. Applicable Documents
   See standard form outline, section 2.

3. Requirements

   3.1 Design

   See standard form outline, section 3.

   3.2 Manufacturing

   See standard form outline, section 4.

   3.3 Testing

   See standard form outline, section 5.

   3.4 Service documentation

   See standard form outline, subsections 6.2 and 6.3.

4. Quality Assurance

   Topics: a. Requirements for a general quality assurance program or manual

   b. Requirements for a particular product-oriented quality assurance plan

   c. Quality assurance document index

   d. Inspection and test plan(s)

   e. Qualifications and certifications

   f. Handling of nonconformances

   g. Auditing for conformance

   h. Required record submittals

5. Preparation for Delivery
   See standard form outline, subsection 6.1.

6. Ordering Data
   Topic: a. Organized numerical data for a particular item or items

APPENDIX E:
COMPONENT DESIGN SPECIFICATION (See Chapter 6)

1. Scope
   Topics: a. Name of equipment
            b. Identification of similar equipment
            c. Equipment functions
            d. System installation
            e. Effects on system operation
2. Applicable Documents
   Same as for the standard form component specification. Should include, if applicable, equipment requirement drawing(s) showing space limitations and connection requirements.
3. Design Criteria
   3.1 Operational design criteria
       See standard form outline, subsection 3.1.
   3.2 Structural design criteria
       See standard form outline, subsection 3.2.
4. Design Features
   Topics: a. Structural features
            b. Mechanical features
            c. Electrical features
   See standard form outline, subsection 3.3.
5. Materials Selection
   See standard form outline, subsection 3.4.
6. Design Documentation
   Topics: a. Reports
            b. Drawings
            c. Operation and maintenance instructions
   See standard form outline, subsection 3.5

APPENDIX F:
COMPONENT TEST SPECIFICATION (See Chapter 6)

1. Scope
   Topics: a. Equipment to be tested
            b. Types of tests to be accomplished (performance, structural, environmental, and endurance or life tests)
            c. Indicate if testing to destruction or failure is included
            d. Indicate if transient as well as steady-state testing is included
2. Applicable Documents
   2.1 Standard statement(s) concerning applicability
   2.2 Design documents

142

List the design specification, the design report, and drawings such as assembly drawings and the installation drawing.

    2.3 Other documents

    2.4 Codes and standards
List in order of precedence, if such an order exists.

3. Test Objectives (Test Criteria)

    3.1 Operational (performance) test objectives (steady-state and transient)

    3.2 Structural test objectives (steady-state and transient)

    3.3 Environmental test objectives

    3.4 Endurance (life) test objectives

4. Test Instrumentation

    Topics: a. Required measurements

             b. Frequency of measurement

             c. Methods of recording readings

             d. Accuracy requirements

             e. Calibration requirements

5. Test Set-Up

    Topics: a. Installation requirements

             b. Operational requirements

6. Test Procedures

    Topics: a. Procedures due to equipment design

             b. Procedures due to test requirements

             c. Order of testing

7. Test Documentation

    Topics: a. Required records and reports

             b. Form of raw data

             c. Calibration records or certifications

             d. Reduced data requirements

             e. Form of final data presentation

APPENDIX G:
SYSTEM DESIGN SPECIFICATION (See Chapter 7)

1. Scope

    Topics: a. System name

             b. What it is used for, or what it is part of

             c. Operational relationship to other systems

             d. Level of criticalness

             e. Cognizant regulatory authorities

2. Applicable Documents

    2.1 Standard statement(s) concerning applicability

    2.2 Company documents

        Topics: a. Drawings controlling system design

b. Specifications or drawings concerning system interfaces

2.3 Codes and standards

List in order of precedence, if such an order exists.

3. Design Requirements

3.1 Design criteria

Topics: a. Applicable codes and standards

b. Details of specific application

3.2 Performance requirements—steady-state

Topics: a. System inputs and outputs

b. Definitions of operating points

c. Operating parameters and ranges

d. Control parameters

e. Set points and limits

3.3 Performance requirements—transient

Topics: a. Rates of change between operating points

b. Primary control parameters

c. Automatic and manual control

3.4 Structural requirements—steady-state

Topics: a. System loadings

b. Expected load combinations

c. Stress and deflection limits

d. Environmental conditions

e. Service life

3.5 Structural requirements—transient

Topics: a. Loading transients

b. Coincidence of steady-state loads and transients

c. Environmental transients

d. Cyclic effects

e. Fatigue life

4. Safety Requirements

4.1 Safety criteria

Topics: a. Applicable codes and standards

b. Details of specific application

4.2 Unsafe conditions

Topics: a. Abnormal operation—steady-state

b. Abnormal operation—transient

4.3 Safety features

Topics: a. Safety system parameters

b. Safety system operation

c. Safety system components

5. Operational Requirements

5.1 Operating modes

Topics: a. Startup

b. Normal operation

       c. Normal shutdown

       d. Abnormal shutdown

5.2 Operating equipment

    Topics: a. Automatic controls

              b. Manual controls

              c. Recorded parameters

              d. Monitored parameters

5.3 Operating locations

    Topics: a. Main controls

              b. Redundant controls

              c. Control arrangement

5.4 Operating personnel

    Topics: a. Operating crew size

              b. Operating crew functions

6. Maintanability Requirements

  6.1 Inspection

     Topics: a. Accessibility

               b. Methods

               c. Equipment

  6.2 Maintenance

     Topics: a. Limitations on removal

               b. Space requirements

               c. Replaceable components

               d. Interchangeability of parts

               e. Limitations on shutdown

  6.3 Removal and replacement

     Topics: a. Limitations on tooling

               b. Limitations on handling

7. Documentation Requirements

  7.1 Drawings

    List all drawing types required for complete definition of the system.

  7.2 Lists

    List all parts, components, and materials lists and sublists required for definition of the system, and their individual requirements.

  7.3 Performance design report

    Describe the contents and level of detail required for this report.

  7.4 Structural design report

    Describe the contents and level of detail, if not specified by codes and standards, required for this report.

  7.5 Operating instructions

    Describe the extent of requirements for preliminary operating procedures.

  7.6 Maintenance instructions

    Describe the extent of requirements for preliminary maintenance instructions.

APPENDIX H:
SYSTEM TEST SPECIFICATION (See Chapter 7)

1. Scope
   Topics: a. System name
             b. What it is used for, or what it is part of
             c. Operational relationship to other systems
             d. Types of tests to be performed
             e. Cognizant supervisory authorities

2. Applicable Documents
   2.1 Standard statements concerning conflicts
   2.2 Design documents
       a. Drawings
       b. Specifications
   2.3 Operational documents
       a. Special manuals (system or subsystems)
       b. Standard manuals (components)
   2.4 Codes and standards
       List in order of precedence, if such an order exists.

3. Acceptance Criteria
   Topics: a. Basis on which acceptable operation is determined
             b. Performance parameters that determine acceptable
                operation

4. Test Conditions
   Topics: a. Steady-state operating limits
             b. Steady-state design points
             c. Additional steady-state points
             d. Transient operating limits
             e. Transient design points
             f. Additional transient points
             g. Switching and transfer points
             h. Safety feature operation

5. Test Instrumentation
   Topics: a. Additional test measurements
             b. Methods of recording readings
             c. Accuracy requirements
             d. Calibration requirements
             e. Monitoring of safety functions

6. Test Procedures
   Topics: a. Procedures for standard system operation
             b. Procedures for special initial system operation
             c. Order of testing

7. Test Documentation

    Topics: a. System acceptance report
                b. General report
                c. Form and disposition of data
                d. Calibration records or certificates
                e. Final settings and adjustments records
                f. Recommendations for modifications to operating procedures

APPENDIX I:
WRITER'S CHECKLIST (See Chapter 14)

1. Have I used the correct format?
2. Have I organized the division into subsections and lists in a logical manner?
3. Have I included all codes and standards required for this type of equipment?
4. If I am using supporting specifications, have I referenced the correct ones?
5. Have I covered all points necessary to completely define the design of this equipment?
6. Have I covered material processing and fabrication in sufficient detail to ensure a product of the desired quality and durability without unnecessarily restricting potential suppliers?
7. Have I included enough tests to assure that the equipment is capable of meeting all design requirements?
8. Have I adequately covered all other requirements for shipping preparation and installation and operational documentation?
9. Are all requirements in the correct locations?
10. Have I complied with all requirements applicable to this type of specification for this equipment and its intended use?

APPENDIX J:
REVIEWER'S CHECKLIST (See Chapter 15)

1. Is the coverage of requirements complete?
2. Are all appropriate company documents used as references or as sources for the contents of this specification?
3. Are all required or appropriate codes and standards applicable to this equipment called for?
4. Are there any other references that should be used?
5. Is there an excess of purely descriptive material?
6. Are unnecessary explanations for requirements included?

7. Are unnecessary discussions of the origins of conditions, such as operating environments or transients, included?

8. Are the requirements adequate for their intended purpose?

9. Are the requirements excessive?

10. Are all special company requirements for this type of equipment included?

11. Are the requirements clear and unambiguous?

12. Has the writer tried to cover a lack of specific requirements with general references to codes and standards?

13. Are any of my comments strictly editorial, and therefore within the editor's area of review?

14. Are all of my comments necessary for improving the specification, or am I simply trying to get the writer to do his or her job my way?

APPENDIX K:
EDITOR'S CHECKLIST (See Chapter 16)

1. Has the typist followed the standard for specifications?

2. Has the correct top-level format been used?

3. Does the table of contents correspond to the first- and second-level section titles?

4. Is the division into second, third, and fourth levels balanced?

5. Is the titling of subsections and paragraphs consistent?

6. Have bottom-line titles been avoided?

7. In parameter lists, have split compound units been avoided?

8. Does the scope section contain standard informational material required by the organization?

9. Does the scope section contain requirements that should appear elsewhere?

10. Is the applicable documents section correctly organized?

11. Is each listing complete?

12. Are all documents listed in this section called out in the requirements sections, and vice versa?

13. Does the applicable documents section contain requirements that should appear elsewhere?

14. Has the limit on levels been observed in the requirements sections?

15. If dependent multiple subsections at the same level have been used, does the specification move up one or more levels immediately afterward?

16. Are all items in lists short enough, and do they contain one requirement only?

17. Have individual requirements been separated?

18. Is the ordering of requirements, modifications, and exceptions consistent?

19. Have sublists been used only for breaking main list items down to their simplest components?

20. Have the requirements been clearly and directly stated?
21. Are multiple-requirement entries correctly used?
22. Are the terms "shall" and "will" correctly used?
23. Are all cross-references correctly numbered?
24. Do subsection titles adequately describe the contents?
25. Do any multiple subsections follow a logical order?
26. Have requirements that belong in other documents of the procurement package been included?
27. Have all grammatical errors been corrected?
28. Have all administrative requirements of the organization been met?

Index

NOW ... *Announcing these other fine books from Prentice-Hall—*

BUILDER'S VEST POCKET REFERENCE BOOK, by William Hornung
$6.95 paperback

ELECTRICIAN'S VEST POCKET REFERENCE BOOK, by Henry B. Hansteen
$6.95 paperback

ELECTRONICS VEST POCKET REFERENCE BOOK, by Harry Thomas
$6.95 paperback

PIPEFITTER'S & PLUMBER'S VEST POCKET REFERENCE BOOK, by George K. Bachmann
$6.95 paperback

MECHANIC'S VEST POCKET REFERENCE BOOK, by John Wolfe, Sc.D., and Everett R. Phelps, Ph.S.
$6.95 paperback

CONSTRUCTION DRAFTER'S VEST POCKET REFERENCE BOOK, by William Hornung
$7.95 paperback

To order these books, just complete the convenient order form below and mail to **Prentice-Hall, Inc., General Publishing Division, Attn. Addison Tredd, Englewood Cliffs, N.J. 07632**

Title	Author	Price*
	Subtotal _____	
	Sales Tax (where applicable) _____	
	Postage & Handling (75¢/book) _____	
	Total $ _____	

Please send me the books listed above. Enclosed is my check ☐ Money order ☐ or, charge my VISA ☐ MasterCard ☐ Account #_____

Credit card expiration date _____

Name _____

Address _____

City _____ State _____ Zip _____

Prices subject to change without notice. Please allow 4 weeks for delivery.